Brain Informatics and Health

Informatics-enabled studies are transforming brain science. New methodologies enhance human interpretive powers when dealing with big data sets increasingly derived from advanced neuro-imaging technologies, including fMRI, PET, MEG, EEG and fNIRS, as well as from other sources like eye-tracking and from wearable, portable, micro and nano devices. New experimental methods, such as in to imaging, deep tissue imaging, opto-genetics and dense-electrode recording are generating massive amounts of brain data at very fine spatial and temporal resolutions. These technologies allow measuring, modeling, managing and mining of multiple forms of big brain data. Brain informatics & health related techniques for analyzing all the data will help achieve a better understanding of human thought, memory, learning, decision-making, emotion, consciousness and social behaviors. These methods also assist in building brain-inspired, human-level wisdom-computing paradigms and technologies, improving the treatment efficacy of mental health and brain disorders.

The Brain Informatics & Health (BIH) book series addresses the computational, cognitive, physiological, biological, physical, ecological and social perspectives of brain informatics as well as topics relating to brain health, mental health and well-being. It also welcomes emerging information technologies, including but not limited to Internet of Things (IoT), cloud computing, big data analytics and interactive knowledge discovery related to brain research. The BIH book series also encourages submissions that explore how advanced computing technologies are applied to and make a difference in various large-scale brain studies and their applications.

The series serves as a central source of reference for brain informatics and computational brain studies. The series aims to publish thorough and cohesive overviews on specific topics in brain informatics and health, as well as works that are larger in scope than survey articles and that will contain more detailed background information. The series also provides a single point of coverage of advanced and timely topics and a forum for topics that may not have reached a level of maturity to warrant a comprehensive textbook.

Daniele Caligiore · Samuele Carli

Simulating the Brain

A Four-Step Method Using Ordinary
Differential Equations and Python

 Springer

Daniele Caligiore
Institute of Cognitive Sciences
and Technologies
Italian National Research Council
Rome, Italy

Samuele Carli ⓘ
Entersys s.r.l.
Padua, Italy

ISSN 2367-1742   ISSN 2367-1750 (electronic)
Brain Informatics and Health
ISBN 978-981-96-2717-2   ISBN 978-981-96-2718-9 (eBook)
https://doi.org/10.1007/978-981-96-2718-9

This Springer imprint is published by the registered company Springer Nature Singapore Pte Ltd.
The registered company address is: 152 Beach Road, #21-01/04 Gateway East, Singapore 189721,
Singapore

If disposing of this product, please recycle the paper.

To Serena and Chiara

—Daniele Caligiore

Fascinating!

—Samuele Carli

Preface

When learning to build brain computational models using the method proposed in this book or other approaches, it is crucial to remember that a model, no matter how detailed or carefully constrained through data, is not reality. It does not matter whether the model represents the functioning of a single neuron or is a "system-level model" that considers the interactions between different areas of the brain and the interplay between the brain, body, and environment. Ultimately, it is still not reality. A model must simplify and capture the essence of reality. Through this simplification, the model becomes a tool that allows us to see and understand aspects of reality that would otherwise be difficult to grasp. A computational model is a tool for formulating new hypotheses about brain functioning or malfunctioning and exploring new paths that are challenging to pursue through reasoning or experiments alone.

Over the years, we have learned to create computational models at various levels of abstraction, ranging from connectionist models to more detailed biological models with spiking neurons and differential equations, as described in this book. We have come to believe that no single method of investigation—whether it be experiments, computer simulations using models, neurophysiology, or brain imaging—is inherently superior to another, nor is there an optimal level of abstraction for creating computational models to study the brain. A computational model that includes more detailed representations of the brain-body-environment system is not necessarily better than a more abstract model that captures general statistical properties and vice versa. The suitability of a model depends on the specific problem studied. A complex model often involves a greater number of free parameters to set, making it unmanageable and difficult to interpret.

Therefore, a model is a tool that can help you think about a problem in new ways compared to traditional "pen and paper" reasoning. However, two reasons make the computational model a unique and, in some ways, special tool compared to other techniques used to study brain, such as empirical experiments or brain imaging. To begin with, a model allows you to *operationalize a theoretical hypothesis* and bring it to life through a *computer simulation*. For instance, you might hypothesize that a lesion in brain area X causes an effect in area Y. With the model, you can verify if this occurs. But there is more. The simulation might reveal that in addition to affecting area Y, the lesion in area X also causes changes in a third brain region, Z, which you had not considered. The simulation results prompt you to reflect on other factors. Computational models, particularly system-level models like those discussed in this book, can integrate and synthesize findings from various approaches and methodologies, such as behavioral experiments, brain imaging, genetic studies, and more. This *integrative capability* is another fundamental aspect that makes computational models a unique and valuable tool for studying the brain through a system-level approach that considers multiple factors.

Thus, the model enables a strongly *interdisciplinary approach* to investigate the brain, supporting an "original multi-methodological language", facilitating synergistic interaction with colleagues from diverse backgrounds. This approach allows for the assimilation and transmission of trans-disciplinary knowledge and promotes *cross-fertilization* between different domains. The interdisciplinary viewpoint allows us to broaden the impact of our research to engage psychologists, neuroscientists, and clinicians aside from modelers. This process fosters innovative perspectives on brain research by promoting collaboration across disciplines like neuroscience, artificial intelligence, psychology, and cognitive science, as well as diverse scientific methods such as (neuro)computational modeling, simulations, machine learning, and experimental psychology. The system-level modeling technique showcased in this book supports this viewpoint.

We have developed an interdisciplinary methodology to design system-level brain models using systems of ordinary differential equations, which are to be solved and analyzed through simple Python scripts. These models could be successfully applied in many contexts: to study healthy and damaged brain functions and hence improve the current understanding of the system-level neural mechanisms underlying brain disorders, but also to discover, through computer simulation, new neural pathways that may be crucial for the emergence of pathologies, as well as the effects of possible new therapies acting on brain actors poorly investigated in traditional research. The main goal of this system-level perspective is to provide an operational hypothesis about the role played by a network of brain areas underlying a behavior: different classes of behaviors are generated by the interplay of various subsets of brain components rather than by specific components in isolation. The proposed method consists of four key steps: (i) designing the model architecture to represent interactions between distinct brain regions; (ii) formulating the system of ordinary differential equations derived from the architecture drawn in (i); (iii) developing a Python script to solve these equations; and (iv) identifying the model free parameters using optimization techniques, such as genetic algorithms or alternative methods, to generate one or more instances that accurately reproduce the target behavior under investigation.

This four-step method has been progressively integrated into the courses we teach at the interdisciplinary "Advanced School in Artificial Intelligence (AS-AI)[1]" directed and co-founded by the first author of this book and promoted by the Institute

---

[1] www.as-ai.org.

of Cognitive Sciences and Technologies of the Italian National Research Council[2] and by AI2Life srl.[3] AS-AI is open to students with any background and actively supports inter-disciplinary collaboration. The idea for this book emerged from this interdisciplinary framework, emphasizing the importance of collaboration and the exchange of ideas across scientific fields. The book is designed for anyone interested in using Python and ordinary differential equations to simulate different aspects of reality, regardless of their background. In addition to explaining the methodology for building system-level brain computational models, the book also includes examples from other dynamic complex systems, such as those in physics and social sciences. While rigorous mathematical proofs of many topics discussed are beyond the scope of this work—since they remain open research areas requiring advanced mathematical expertise—the essential elements needed for critical analysis and self-assessment are presented in a simplified and practical manner. For those interested in delving deeper, extensive references are provided. This textbook is a comprehensive resource, providing everything needed to learn from the ground up. It covers essential mathematical concepts—ranging from foundational topics for beginners to insights on advanced subjects for more experienced readers—as well as fundamental computing and collaboration tools indispensable for interdisciplinary, team-based research. Additionally, it introduces the basics of Python programming and equips readers with the skills to design, simulate, visualize, and interpret models of the brain and other complex systems. These abilities are honed through numerous hands-on examples, explained step-by-step.

Keywords: Brain simulation, Digital twin, Hands-on examples, Interdisciplinary, Network neuroscience, Numerical models, Python, Ordinary differential equations, System-level modeling.

Acknowledgments for Daniele Caligiore

I express my gratitude to my colleagues at the Italian National Research Council—Institute of Cognitive Sciences and Technologies and AI2Life srl, and the lecturers and students at the Advanced School in Artificial Intelligence. My heartfelt thanks go to Serena, Chiara, Carla, Claudio, Francesca, Maria Enza, Mauro, Miriam, Mirco, Nella, Nello, Paola, and Salvatore for their support and encouragement.

Acknowledgments for Samuele Carli

Mentioning all the incredible friends and collaborators that accompanied me during this journey, and everything they did with me and for me, would require a book of its

---

[2] www.istc.cnr.it.

[3] www.ai2life.com.

own. You know who you are, and I'm sure you also know how grateful I am. Thank you! Osobitné poďakovanie patrí mojej úžasnej manželke Ine, e grazie a mamma e babbo!

Rome, Italy   Daniele Caligiore
Padua, Italy   Samuele Carli

Competing Interests The authors have no competing interests to declare that are relevant to the content of this manuscript.

Contents

About the Authors

Daniele Caligiore is a Research Director at the Institute of Cognitive Sciences and Technologies (ISTC) of the Italian National Research Council (CNR), where he coordinates the Computational and Translational Neuroscience Laboratory (CTNLab). This interdisciplinary research group leverages artificial intelligence (AI) and computational neuroscience to study brain function and behavior. The CTNLab ultimate goal is to discover new clinical applications and therapies for neurodegenerative diseases. He is the director and co-founder of the Advanced School in Artificial Intelligence (AS-AI), a postgraduate institution supported by ISTC, dedicated to the interdisciplinary study and application of AI. Additionally, he teaches "Artificial Intelligence Systems in Social Contexts" at the Libera Università Maria Santissima Assunta (LUMSA) in Rome. Daniele earned his Master's Degree in Electronic Engineering from the University of Catania in 2003 and a PhD in Biomedical Engineering from the University Campus Bio-Medico of Rome in 2011. He has been a visiting scholar at the University of Plymouth, the University of Bologna, and the University of Southern California. With over 90 scientific articles published in international journals, conference proceedings, and book chapters, Daniele has also contributed to numerous international research projects in artificial intelligence and neuroscience. He serves on the Scientific Secretariat of the CNR Commission for Ethics and Integrity in Research and is an expert evaluator of research project proposals in AI and neuroscience for the Italian Ministry of University and Research (MUR) and the European Commission. Additionally, he is a founding member of AI2Life, a spin-off of ISTC-CNR, which aims to promote the use of AI to foster social development and individual well-being. Daniele is the author of interdisciplinary books, including "*IA istruzioni per l'uso*" and "*Curarsi con l'Intelligenza Artificiale*", both published by il Mulino, and "*Healing with Artificial Intelligence*", published by Routledge Taylor & Francis.

Samuele Carli is a seasoned Senior Developer and Entrepreneur with over 25 years of professional experience. Throughout his career, he has contributed to a diverse array of projects across scientific research, industry, and business domains, taking on roles ranging from developer to project manager and coordinator. He holds a

Master of Science degree in Computer Science from Università degli Studi di Firenze, specializing in numerical analysis, computer simulation, and artificial intelligence. His expertise includes work on machine learning applications for data analysis at CERN (the European Organization for Nuclear Research). Currently, he collaborates as a researcher with CTNLab and ISTC-CNR, focusing on brain modeling and simulation in the context of neurodegenerative diseases. In addition to his research endeavors, Samuele teaches courses on numerical optimization at the Advanced School in Artificial Intelligence. He is also the co-founder and Chief Technology Officer of Entersys S.r.l., a company dedicated to small enterprise management and the development of AI-driven and data-centric solutions for enterprise resource planning and management.

List of Figures

Chapter 1
Essential Computing Tools

This chapter presents the fundamental technical concepts essential for any computational project. It addresses key aspects of software development, effective group collaboration, and the Python programming language, providing a comprehensive introduction to the realm of software development.

1.1 Programming Styles and Development Methodologies

1.1.1 Procedural Programming

Procedural programming is a programming paradigm based on the concept of the procedure call. Procedures, in the past also known as routines, subroutines, or functions, consist of a series of computational steps to be carried out. Any given procedural program is made up of one or more procedures. In a procedural program, execution is *sequential*: instructions are executed in a linear, step-by-step manner, and the program flow is entirely controlled by the flow statement of the language, such as 'if-else' (or even 'go-to's in less abstract and older languages) constructs and function calls. Functions contribute to making portions of code reusable and manageable. The linear and (oftentimes only seemingly) straightforward nature of procedural programming can make it easier to understand and implement an algorithm or a model, especially in small projects.

However, as the size of the codebase grows, this approach can quickly become difficult to manage. The reusability of functions tends to diminish as the project grows, because they usually rely on abstractions that are only pertinent in the scope of that particular aspect of the project. This in turn can lead to excessive code duplication, which negatively affects the maintainability of the whole codebase.

Procedural code tends to be leaner and more efficient than an equivalent object oriented one, therefore it can be the recommended approach in such situations where

© The Author(s), under exclusive license to Springer Nature Singapore Pte Ltd. 2025
D. Caligiore and S. Carli, *Simulating the Brain*, Brain Informatics and Health,
https://doi.org/10.1007/978-981-96-2718-9_1

performance is of critical importance. In procedural programming, states and behaviours are completely separated; this is of course not a bad thing it itself, but unless the developers are being rigorous and meticulous with their abstractions and representations, as the code base organically grows and functionalities are added, they tend to get mixed in ways that can become difficult to understand later on when the code needs to be reworked or audited. Most computer languages allow writing programs procedurally.

1.1.2 Object-Oriented Programming

Object-Oriented Programming (OOP) is a programming paradigm that uses objects and classes to organize code into *reusable and related units*. In particular, a class can be seen as a blueprint (or template) to create objects. A class is an abstraction of some entity which has a state (data represented as attributes) and a behaviour (methods which operate on the object). An object is therefore a particular instance of a class, defined by its current state (the value of its attributes). All objects of the same class are completely independent, but they all have the same attributes (each with their own different values) and can all perform the same actions. For example, one could define the class of Car: all have the same attributes (make, model, year, color, odometer) and all can start_engine, stop_engine, go_to, park. A yellow fiat 500 from 1969 with 100000 km on the odometer is a particular instance (object) of the class Car, which can potentially be instructed to start the engine and go somewhere.

The first important difference between OOP and procedural programming is that, through classes, data and its handlers are represented and kept together; furthermore the class implicitly describes the abstraction that is being done, hence gives directly a meaning and purpose to the data which is easier for the programmer to find and understand. This bundling of data and methods is called *encapsulation*. In principle, all the attributes and internal workings of a class should be private and not accessible by other objects; a class should instead offer a public interface which includes methods to read and modify its state and trigger behaviours. This encapsulation allows to completely change the inner working of a class, including the representation of the data and so on, without changing its interface: therefore, without having to update any of the code that uses this class. This is already a great advantage over the procedural programming approach, where one must potentially search through and modify the whole codebase to change some data representation detail.

Another important feature of classes is *inheritance*: a class can inherit and extend (or modify) the attributes and methods of another class. This allows for greater code reuse, and also allows to change or extend the behaviour of existing code without modifying it (for example, a third-party library can provide most of the implementation, and one only modifies the behaviour of some methods or adds some attributes while reusing most of the provided code). For example, we could define a class Van that inherits from the Car we defined earlier, but adds a cargo_volume attribute and the "load" and "unload" methods. An instance of Van therefore will have all the methods and attributes defined in Cars, plus the new ones we just added.

This brings us directly to the concept of *polymorphism*: since a Van has the same methods and attributes of a Car, in a sense a Van *is* a Car: an object instantiating Van can be used instead of a Car, in exactly the same way the Car was used. A piece of code that has to check how many kilometers a vehicle was driven, can expect to be passed an object of type Car. But at runtime, it could be passed an object of type Van and everything would work just the same, without need to change any code! Furthermore, Van could be changing the Car implementation of start_engine and stop_engine, since the Van, having a different kind of engine, needs a different procedure. Notice how all this changes to the program's behaviour can happen without having to touch any of the code that uses the Car class!

Summarizing, Object Oriented code allows to develop *modular* code, where classes can be easily reused and extended in other contexts; this also makes the code more *reusable* (of course if the abstractions have been chosen carefully, that is). OO also improves *maintainability*, since individual components of the system can be upgraded and modelled to fit any new environment pretty much independently. It is important to study and employ *design patterns*, which are proven solutions to common problems, leading to more robust and scalable software design [4–6]. Of course, OOP can also introduce some complexity in a codebase, especially in very small codebases where the overhead of defining classes and objects may be unnecessary. It is our experience, however, that very small codebases rarely tend to remain such, as projects grows in requirements and functionality, hence in general it is a really good idea to always consider an object oriented approach, and eventually refactor part of the code to become procedural in the (extremely rare) cases where the object orientation actually hinders the performances of a critical part of a system. The initial overhead in the design time of a piece of OO software is nearly always compensated by the clarity and maintainability of the code in the long run.

1.1.3 Test Driven Development

Test-Driven Development (TDD) is a software development methodology where tests are written before writing the actual code. TDD is usually implemented around the following cycle:

- Write a Test: Before any functional code is written, a test is created that specifies and validates what the code will do.
- Run All Tests: Running all tests ensures the new test fails, which confirms that the test is valid and the feature isn't already correctly implemented.
- Write the Code: Write the minimal code necessary to pass the test.
- Run All Tests: Ensure all tests pass. If they don't, the new code is adjusted until they do.

- Refactor Code: Clean up the code while ensuring that all tests still pass. Refactoring improves code readability and maintainability without changing its behavior.

Employing a technique such as TDD has some distinct advantages: writing tests first ensures that the code is testable, meets the requirements and reduces the likelihood of defects. The design of the code ends up being better, since testability also implies simplicity, granularity and effective interfaces. Ensuring the presence of tests also adds a degree of regression prevention, since future refactorings which introduce nonconformities will break the existing tests. Tests are also implicitly a form of documentation, since they specify exactly what the tested code is supposed to do. An existing test suite also encourages code refactoring and maintainance, since tests act as an easy safety net to find quickly any possible mistake in the code changes.

Writing tests before code can however be seen as a slowdown in development (even though it *is* a great investment in the future of the project!), and can be challenging especially in some initial prototyping stages when the project is supposed to be moving and changing fast. However, experience dictates that as soon as a prototype semes to be working, it magically becomes production code: tests are never written, the code is never refactored, and a few months down the line, disaster strikes. The initial overhead of writing tests together with code at least ensures that some safety net will be present when this *will* happen... It is also important not to fall in the pitfall of putting too much emphasis on test for the sake of writing tests: there is little value in testing lots of trivial code paths while missing out on the broader integration issues. Tests should cover the common pitfalls and corner cases (did you mistakenly shift your index by 1? is your function crashing with an empty input?) and most importantly, it should also test the interoperation of components around the corner cases. A few tests that actually look for the correct behavior in a functional and theoretical sense is more valuable than many tests which only look at trivial functions. Tests should be simple to understand and loosely coupled with the code, to allow for easy refactoring and maintainance of both the tests and the code. Writing tests after code, ignoring the "test-first" principle, can lead to missed requirements and less effective tests; however a tight cycle of code-tests-code-tests is much, much better than no tests at all.

1.1.4 Continuous Integration/Continuos Delivery

In modern software development, Continuous Integration (CI) and Continuous Delivery (CD) have become fundamental practices that aim to enhance the efficiency, reliability, and speed of delivering software. *Continuous Integration* is a development practice where developers integrate code into a shared repository frequently, usually several times a day. Each integration is verified by an automated build and automated tests to detect integration errors as quickly as possible. Integrating development work in such small steps offers several advantages: in primis, every developer in the team can profit instantly of any refactoring, new test or new functionality that is integrated in the projet. In turn, this means that eventual code conflits will get resolved very quickly as soon as they are discovered, when they are still small and manageable. Merging branches that diverged for months and thousands of changes is a much more difficult and costly endeavour, and is also more likely to introduced bugs because the merge happens far away in time with respect to the development, and small details are most likely been forgotten by every developer that already moved on to another task. The tight feedback loop also ensures that issues are identified and fixed quickly, leading to a higer quality of the codebase averall. Considering that one should always have an automated test framework in place, the biggest cost associated with CI is, arguably, the argument that developing in such small steps can be slower for the programmer, which may have to do extra work to ensure that each intermediate step in the development does not break the code for everyone else. At the same time, though, it could be argued that this approach produced better, clearer and simpler code, and the impression of slower development in reality repays itself very quickly in terms of code quality.

Continuous Delivery is an extension of Continuous Integration that aims to ensure that the software can be reliably released to production at any time. It involves deploying every change that passes the automated tests to a staging environment and preparing it for release to production. After CI processes are complete, the build is automatically deployed to a staging environment where further automated integration, performance and security tests are performed. Once all tests are passed, the release is tagged as ready for deployment and, possibly after a manual approval phase, it goes directly in production. This approach shortens the feature-time-to-use to the bare minimum, so the new features and bug fixes can be profited by the end user right away. Very small, incremental releases also minimize the upgrade risk. The additional automated testing required to maintain this approach elevates the confidence in the correctness and quality of the whole codebase, while the short deployment cycles ensure a quick customer feedback that can more effectively lead the developers in the right direction. This approach however adds complexity to the project, and can be quite resource intensive.

In conclusion, CI/CD are powerful practices that can significantly enhance the efficiency, quality and speed of software development, ensure higer quality standards by enforcing extensive automated testing and deliver new features and fixes with the shortest possible delay. By automating key processes, fostering collaboration, and ensuring a continuous feedback loop, CI/CD helps teams deliver reliable software more rapidly and respond effectively to changing requirements. While there are challenges and costs associated with implementing CI/CD, the benefits far outweigh the drawbacks, making it an essential part of modern software engineering.

TDD and CI/CD are essential tools to be applied also (if not especially) in the context of writing code for research purposes. They ensure a better level of correctness and understanding, which in turn gives the researcher much more confidence in the obtained results. Furthermore, as reproducibility is a fundamental aspect of modern research, leaving behind tested and well structured, updatable code is a strong quality and credibility aspect that all research code should deliver [7–9].

1.1.5   Software, Data, and Their Fight Against Time

Software development is an iterative, never-ending process that must adapt to the constantly changing technological landscape. While the functionality itself of a piece of code may not need to change for a very long time, the environment in which this piece of code will run is bound to change as the world around it evolves: operating systems, libraries, standards, networks, processor architectures, computing models, security vulnerabilities... Eventually, any piece of code will have to be kept up to date with its environment or it will cease working; software must be constantly maintained to survive. Unfortunately this is also true for data: without the running software necessary to interpret it, also data "rots" in time as the knowledge about it vanishes. Especially in the scientific world, where studies' reproducibility and scrutinability in the long term are essential, employing tools like virtual machine snapshots, containers, emulation, and source code archiving is essential to ensure that software remains functional and valuable for future generations. Having proper, clean, correct and complete documentation of data structures and algorithms is also an essential tool that helps both the developer of today, and the one of tomorrow. Data should be labeled as clearly as possible (what is stored in this `data_new` directory anyway? And why is there a `new_data_new_better` which contains gigabytes of stuff that is not used by any of the code?). Be careful and don't be lazy when developing; a good documentation and labeling strategy may take a minute longer today, but will surely save you many headaches already while working now and a whole lot of time tomorrow, when someone else may have to understand it all again.

1.2   Versioning and Collaborating

In the world of software development (and science even more so), collaboration and effective project management are crucial. This is where version control systems (VCS) like Git, Mercurial (hg), and Subversion (SVN) come into play. These tools are indispensable for managing changes to source code over time and facilitating teamwork. VCSs are fundamental for:

- Tracking changes: with a VCS, developers can track changes in the codebase. Every modification by any developer is logged, and in principle all changes can (and should) be audited, code reviewed and tested (automatically by a test suite!) before being integrated in the main branch. This historical record is invaluable for debugging (individual changes can be reverted, reapplied, moved to other branches for further testing etc.) and to keep track of the evolution of a project (what was done, when, by whom...);
- Collaboration: In a team setting (but can also happen when working alone!), multiple developers can work in parallel on different parts of the project, and their work can be integrated (merged) automatically by the VCS system, which will also track and help in resolving conflicting changes;

- Backup and Recovery: a VCS is intrinsically also a robust backup system: each previous version (each change in fact!) of the project can be restored, examined and compared. If something is accidentally deleted or modified, the previous state can be restored; most VCS also work in a distributed manner, allowing naturally and easily to have multiple copies of the project in different places which is a great way of having backups;
- Experimentation: Features can be implemented in separate branches, experimented on and evaluated before becoming part of the main project; the freedom of experimentation without breaking the project for everyone else is invaluable to progress!

Git is the most widely used version control system today. Created by Linus Torvalds in 2005, Git is known for its speed, flexibility, and distributed nature. Each developer has a complete copy of the repository, including its history, which enhances performance and reliability. Branches allow team members to work on different features or bug fixes independently. Once their work is complete, they can merge their changes back into the main branch. This workflow minimizes disruptions and ensures that the main codebase remains stable. When multiple developers make changes to the same part of a codebase, conflicts can occur. Git detects these conflicts and alerts the developers, who can then resolve the issues before merging. This process prevents accidental overwrites and ensures that all contributions are integrated smoothly. Code reviews are usually done through pull requests or similar mechanisms. Team members can review each other's code before it is merged, providing feedback and catching potential issues early. This collaborative review process improves code quality and fosters knowledge sharing.

1.2.1 A Git Primer

Let's create a local git repository and interact with it a little bit. We'll do that in a temporary directory:

```
1   export TERM=dumb
2   rm -rf /tmp/test_repo
3   mkdir -p /tmp/test_repo
4   cd /tmp/test_repo
5   git init .
```

```
1   hint: Using 'master' as the name for the initial branch.
    ↪    This default branch name
2   hint: is subject to change. To configure the initial
    ↪    branch name to use in all
3   hint: of your new repositories, which will suppress this
    ↪    warning, call:
4   hint:
```

```
5  hint:              git config --global init.defaultBranch
   ↪   <name>
6  hint:
7  hint: Names commonly chosen instead of 'master' are
   ↪   'main', 'trunk' and
8  hint: 'development'. The just-created branch can be
   ↪   renamed via this command:
9  hint:
10  hint:              git branch -m <name>
11  Initialized empty Git repository in /tmp/test_repo/.git/
```

The repository has only one empty branch, which got the default name 'master'. As the help message said, this can be changed. In fact, all branches are democratically equals as far as git is concerned, and branch naming is just a convention that one can choose at will. We can configure the author's name and email for the current repository, and create a file:

```
1  git config user.name "Arthur the Author "
2  git config user.email "Arthur@author.commit"
3  echo 'hello' > a_file
4  git status
```

```
1  On branch master
2
3  No commits yet
4
5  Untracked files:
6    (use "git add <file>..." to include in what will be
   ↪   committed)
7          a_file
8
9  nothing added to commit but untracked files present (use
   ↪   "git add" to track)
```

The status report shows that there is a file which git is not tracking (yet). Untra-caked files can be cleaned up or ignored (see the documentation of git clean and .gitignore). In our case, we want to add this file to the git index:

```
1  git add a_file
2  git status
```

```
1   On branch master
2
3   No commits yet
4
5   Changes to be committed:
6     (use "git rm --cached <file>..." to unstage)
7           new file:   a_file
```

Now the file is tracked, but is new and its changes were never commetted. Let's create a commit:

```
1   git commit -m "Added a_file to the repository!"
```

```
1   [master (root-commit) 585b128] Added a_file to the
    ↪  repository!
2    1 file changed, 1 insertion(+)
3    create mode 100644 a_file
```

```
1   git log
```

```
1   commit 585b12873affd8714b07fd4bda0bfa4ca095547a (HEAD ->
    ↪  master)
2   Author: Arthur the Author <Arthur@author.commit>
3   Date:   Wed Jul 31 11:03:23 2024 +0200
4
5       Added a_file to the repository!
```

```
1   git status
```

```
1   On branch master
2   .nothing to commit, working tree clean
```

We just created the first commit of this repository! The log shows when and by whom the commit was done, and we see that the status is 'clean', meaning, nothing has changed with respect to what is committed.

```
1  echo "let's add another line to a_file" >> a_file
2  git status
```

```
1  On branch master
2  Changes not staged for commit:
3    (use "git add <file>..." to update what will be
   ↪  committed)
4    (use "git restore <file>..." to discard changes in
   ↪  working directory)
5        modified:   a_file
6
7  no changes added to commit (use "git add" and/or "git
   ↪  commit -a")
```

After changing something, the status described what was changed, and git can produce a diff (also called patch) which shows exactly what changed.

```
1  git diff
```

```
1  diff --git a/a_file b/a_file
2  index ce01362..ca8b0d9 100644
3  --- a/a_file
4  +++ b/a_file
5  @@ -1 +1,2 @@
6   hello
7  +let's add another line to a_file
```

```
1  git add a_file
2  git commit -m "Added a line to a_file"
3  git log
```

```
1   [master d736fda] Added a line to a_file
2    1 file changed, 1 insertion(+)
3   commit d736fda4b16186e4cd431644591b13bb86a77dcc (HEAD ->
   ↪  master)
4   Author: Arthur the Author <Arthur@author.commit>
5   Date:   Wed Jul 31 11:03:23 2024 +0200
6
7       Added a line to a_file
8
9   commit 585b12873affd8714b07fd4bda0bfa4ca095547a
10  Author: Arthur the Author <Arthur@author.commit>
```

```
11   Date:    Wed Jul 31 11:03:23 2024 +0200
12
13       Added a_file to the repository!
```

We added and committed the new change. Now the history has two commits. Say we now need to experiment on something which we are not sure we want to integrate in the main branch. We can create a new feature branch and do some work in it:

```
1   git checkout -b feature_branch
2   git branch
```

```
1   Switched to a new branch 'feature_branch'
2   * feature_branch
3     master
```

```
1   echo -e "This was done in the feature_branch\nlet's add
    ↪   another line to a_file" > a_file
2   git status
3   git diff
```

```
1   On branch feature_branch
2   Changes not staged for commit:
3     (use "git add <file>..." to update what will be
      ↪   committed)
4     (use "git restore <file>..." to discard changes in
      ↪   working directory)
5         modified:   a_file
6
7   no changes added to commit (use "git add" and/or "git
    ↪   commit -a")
8   diff --git a/a_file b/a_file
9   index ca8b0d9..d203abf 100644
10  --- a/a_file
11  +++ b/a_file
12  @@ -1,2 +1,2 @@
13  -hello
14  +This was done in the feature_branch
15   let's add another line to a_file
```

```
1   git add a_file
2   git commit -m "Changed the first line"
3   git log
```

```
1    [feature_branch e4b4c4d] Changed the first line
2     1 file changed, 1 insertion(+), 1 deletion(-)
3    commit e4b4c4d9a681c8dd5a2eafe4a3200b66390560b5 (HEAD ->
     ↪   feature_branch)
4    Author: Arthur the Author <Arthur@author.commit>
5    Date:   Wed Jul 31 11:03:24 2024 +0200
6
7        Changed the first line
8
9    commit d736fda4b16186e4cd431644591b13bb86a77dcc (master)
10   Author: Arthur the Author <Arthur@author.commit>
11   Date:   Wed Jul 31 11:03:23 2024 +0200
12
13       Added a line to a_file
14
15   commit 585b12873affd8714b07fd4bda0bfa4ca095547a
16   Author: Arthur the Author <Arthur@author.commit>
17   Date:   Wed Jul 31 11:03:23 2024 +0200
18
19       Added a_file to the repository!
```

Now the feature branch has one commit that represents the "side work" we have done. In the meantime, work on the master branch continued:

```
1    git checkout master
2    git log
```

```
1    Switched to branch 'master'
2    commit d736fda4b16186e4cd431644591b13bb86a77dcc (HEAD ->
     ↪   master)
3    Author: Arthur the Author <Arthur@author.commit>
4    Date:   Wed Jul 31 11:03:23 2024 +0200
5
6        Added a line to a_file
7
8    commit 585b12873affd8714b07fd4bda0bfa4ca095547a
9    Author: Arthur the Author <Arthur@author.commit>
10   Date:   Wed Jul 31 11:03:23 2024 +0200
11
12       Added a_file to the repository!
```

```
1    cat a_file
```

```
1  hello
2  let's add another line to a_file
```

```
1  echo -e "adding more work done to the file" >> a_file
2  git status
3  git diff
4  git add a_file
5  git commit -m "More development work done on master,
   ↪  changed the second line"
```

```
1   On branch master
2   Changes not staged for commit:
3     (use "git add <file>..." to update what will be
    ↪  committed)
4     (use "git restore <file>..." to discard changes in
    ↪  working directory)
5           modified:   a_file
6
7   no changes added to commit (use "git add" and/or "git
    ↪  commit -a")
8   diff --git a/a_file b/a_file
9   index ca8b0d9..40516c5 100644
10  --- a/a_file
11  +++ b/a_file
12  @@ -1,2 +1,3 @@
13   hello
14   let's add another line to a_file
15  +adding more work done to the file
16  [master 7316a61] More development work done on master,
    ↪  changed the second line
17   1 file changed, 1 insertion(+)
```

```
1  git log
```

```
1  commit 7316a61b82ea0a8a62965635f467f78eadb164f0 (HEAD ->
   ↪  master)
2  Author: Arthur the Author <Arthur@author.commit>
3  Date:   Wed Jul 31 11:03:25 2024 +0200
4
5      More development work done on master, changed the
    ↪  second line
6
7  commit d736fda4b16186e4cd431644591b13bb86a77dcc
8  Author: Arthur the Author <Arthur@author.commit>
```

```
9   Date:    Wed Jul 31 11:03:23 2024 +0200
10
11       Added a line to a_file
12
13   commit 585b12873affd8714b07fd4bda0bfa4ca095547a
14   Author: Arthur the Author <Arthur@author.commit>
15   Date:    Wed Jul 31 11:03:23 2024 +0200
16
17       Added a_file to the repository!
```

```
1   cat a_file
```

```
1   hello
2   let's add another line to a_file
3   adding more work done to the file
```

We added one more commit in the master branch. In the meantime, it is finally decided that our feature branch should be merge in:

```
1   git merge --no-edit feature_branch
```

```
1   Auto-merging a_file
2   Merge made by the 'ort' strategy.
3    a_file | 2 +-
4    1 file changed, 1 insertion(+), 1 deletion(-)
```

In this case, git could automatically figure out that the changes made to the file were not conflicting, and managet to marge the work in:

```
1   git status
2   git log
```

```
1   On branch master
2   nothing to commit, working tree clean
3   commit 52063ca9c7af438de0399dd582ed3b3c48cacde2 (HEAD ->
    ↪   master)
4   Merge: 7316a61 e4b4c4d
5   Author: Arthur the Author <Arthur@author.commit>
6   Date:    Wed Jul 31 11:03:25 2024 +0200
```

```
7
8      Merge branch 'feature_branch'
9
10 commit 7316a61b82ea0a8a62965635f467f78eadb164f0
11 Author: Arthur the Author <Arthur@author.commit>
12 Date:   Wed Jul 31 11:03:25 2024 +0200
13
14      More development work done on master, changed the
       ↪  second line
15
16 commit e4b4c4d9a681c8dd5a2eafe4a3200b66390560b5
   ↪  (feature_branch)
17 Author: Arthur the Author <Arthur@author.commit>
18 Date:   Wed Jul 31 11:03:24 2024 +0200
19
20      Changed the first line
21
22 commit d736fda4b16186e4cd431644591b13bb86a77dcc
23 Author: Arthur the Author <Arthur@author.commit>
24 Date:   Wed Jul 31 11:03:23 2024 +0200
25
26      Added a line to a_file
27
28 commit 585b12873affd8714b07fd4bda0bfa4ca095547a
29 Author: Arthur the Author <Arthur@author.commit>
30 Date:   Wed Jul 31 11:03:23 2024 +0200
31
32      Added a_file to the repository!
```

```
1  cat a_file
```

```
1  This was done in the feature_branch
2  let's add another line to a_file
3  adding more work done to the file
```

In some cases (for example, if the same line is changed in different ways in the two branches, or the context changes enough that git cannot figure it out), the merge will fail with a conflict error. In such cases, the files that couldn't be merge will contain the conflicting blocks coming from both branches, and it's up to us to choose which block is the correct one (or to do something entirely different if needed). Once the conflict is resolved, the modified files can be added and `git merge --continue` will allow us to finish up the work we started.

With git, all repositories are equal. You can pull and push commit history to and from any branch of any remote repository. Usually, there is agreement on which is the main (or reference) repository, that will be hosted on some server accessible to all developers; each of them will then `clone` the main repository locally, work, commit their work, and then push to the main repository when appropriate depending on the chosen workflow. Sometimes, developers will each have two (or more) repositories: some on their local machines to work, and one on the server, which can be used to share their work with others (by means of pull request or other mechanisms).

1.2.2 Development Approaches: Branching and Merging Strategies

In the realm of collaborative development, particularly in software projects, several strategies have proven to be highly effective for ensuring code quality, maintaining consistency, and fostering teamwork.

Code Review is a systematic examination of code by one or more developers other than the author, aimed at finding and fixing bugs, improving code quality, and sharing knowledge. The most important outcome of code review is the sharing of knowledge: reviewers will give important feedback about code quality, coding ideas, performance, correctness, testing, style and so on which, in the long run, will elevate every developer's competence in all aspects of their professional life. Also, most mistakes will be cought early by the reviewers, improving the overall code quality. Last but not least, the reviewing process will slowly bring consistency across the development style of all developers, which in turn will make for better code quality and easier life for everyone involved. It is however of paramount importance that all the parties involved understand that this is not a power game: the aim is to produce better code and to help each other being better developers, not demonstrating one's "superiority". Comments on reviews must be kept constructive and focused on the code, through small, incremental reviews which provide specific and actionable feedback. The objective is always to grow together, no matter your seniority level!

Pull Request is a method of submitting contributions to a project. It allows developers to notify team members that they have completed a feature or fix and would like it to be reviewed and merged into the main codebase. With this process, a developer works indipendently on a feature, and submits the results for review and integration once they are ready. The project integrator will review, eventually go throguh a cycle of feedbacks and improvements, and will merge in the changes when the work is deemed ready. This approach has the benefit of isolating the work on features, which will not affect the main codebase in favor of parallelism, and facilitates the code review process, which is fundamental. However, when work on features done in parallel by multiple developers starts to diverge significantly (for example with big features that take a long time, or refactoring work), the merging process becomes more and more difficult, error-prone and time-consuming. Also, the project integrator may not always have the specific knowledge of the developer needed to resolve merge conflicts, which can introduce bugs or require more time than necessary. The pull requests strategy is therefore better used in a continuous integration context, where branches are as short lived as possible and integrated frequently; this approach keeps all the organizational advantages of code review and pull requests, while also providing the main advantages of continuous integration, which are smaller and easier merge conflicts, shorter feature delivery times, developer synchronization and easy refactoring.

Merging is the process of integrating changes from one branch into another, typically from a feature branch into the main branch. A merge can be fast-forward, when all commits of the branch being merged in were on top of the last commit of the main branch, or it may involve a three-way merge, when the branch histories diverged after a common commit ancestor. The three-way merge may requre manual work when the merging algorithm cannot figure out how to resolve conflicting changes (in general, when the same line in two files was changed in different ways in the two branches). This approach (as an alternative to rebasing, described later) has the benefit of keeping the full commit history, so that there is full transparency on how the work was done, and in general it is easier to identify which change introduced a bug, making it also easier to subsequently fix.

Rebasing is an alternative to merging, which involves moving or combining a sequence of commits to a new base commit. This approach ensures a linear history, and delegates the resolution of conflicts to the feature's developer, which must rebase and resolve conflicts by rebasing before submitting the most recent work. This however becomes a cumbersome approach when there are many developers working in parallel: every developer in principle should rebase their work every time a change is merged into the main branch, stealing precious time and efforts that could be dedicated to developing new features, and creating stale times during which the integrator has to wait for all developers to rebase after he merged in a feature, and may require specific strategies to minimize unnecessary work. Also, rebasing (and hence, rewriting history) is a delicate process that can lead to mistakes and introduce bugs if not done carefully.

1.3   The Python Programming Language

Python is widely regarded as one of the best languages for implementing scientific simulation code due to its simplicity, readability, and extensive ecosystem of libraries. Its easy-to-understand syntax reduces the complexity of translating mathematical models into code, allowing scientists to focus on problem-solving rather than programming intricacies. Additionally, Python's robust libraries, such as NumPy for numerical computations, SciPy for advanced scientific calculations, and Matplotlib for data visualization, provide powerful tools for developing and analyzing simulations efficiently. The strong community support and constant development of new packages further enhance Python's capabilities, making it an indispensable tool for scientific research and simulation. There are (rare) cases where Python may not be performant enough for the task at hand, but it is easy to interface and extend Python programs with lower level languages like C, C++ and Rust; performance critical parts of the any program can be implemented in such languages while Python glues everything together. Python is usually a great first choice to carry out any of the tasks needed by a scientific project.

1.3.1   Incremental Development

When developing using compiled languages such as Java or C/C++, usually the development cycle goes like this:

1. Write the minimal boilerplate code necessary for the language to do anything at all. For example:

```
#include <iostream>
int main(int argc, char*[] argv) { return 0; }
```

2. Write a bit of code that actually **does** something
3. Compile
4. Look at the compilation errors
5. Yell at the semicolon you forgot
6. Edit
7. Save
8. Compile
9. Look at the compilation errors
10. Try to figure out what you did wrong
11. Edit
12. Save
13. Go back to step 3.

   And what if you would need to quickly test the difference between two alternative approaches in doing something trivial? Most probably you would have to recompile the whole project multiple times; not only you'd have to wait, but the time spent waiting for the compilation breaks your concentration and disrupts your workflow; or ever worse, you might have to write lots of boilerplate code just to test and better understand some small snippet of code. One of the greatest Python advantages is exactly its interactive nature. At any time, you can launch your program in an interactive shell, and test tiny fragments of code on-the-fly, instantly, to get important feedback on your ideas without needing additional time-consuming steps. While interacting with the python interpreter, the program's state is not lost: you can recover from runtime errors, examine the state of the program much like if you were using a debugger, understand what might have gone wrong and continue with the execution of the program. This process can boost development speed by orders of magnitude, when compared with the debug/compile cycle!

More importantly, interacting with the interpreter can lead development in a better direction that produces higher quality code. Imagine the kind of conversation that you would have with someone so far away that there was a transmission delay of one minute or more. Now imagine interacting with someone in the same room as you. You wouldn't just be having the same conversation faster: you would be having an entirely different conversation! Having the possibility to interact with the interpreter in real-time is much like having a conversation face-to-face: you can test your ideas as you're writing them; this kind of instant feedback has a dramatic effect on the code you are writing as it does on a conversation. Use this feature to your advantage!

1.3.2 Alternatives and Complements

It can happen that scientific code requires some computationally intensive tasks where the interpreted nature of python becomes a burden that slows down some critical parts of the code. While it's always better to try other approaches first, such as leveraging the native C implementation of many Python libraries like numpy, pandas etc., a portion of your Python code can be rewritten in a more performant language without losing the advantages that Python brings to everything else.

As the Python interpreter itself is implemented in C, when there are explicit and easy-to-identify bottlenecks (like tight loops in a function which does some local computation) it is usually worth trying Cython, which is an extension of python that allows to write Python-like syntax that gets compiled into C code at runtime, leading to substantial performance improvements. This is particularly useful for loops and mathematical operations that require optimization, and being well integrated in the Python syntax, it is usually fast and easy to try. Alternatively, one can always leverage the Python C API and write C/C++ python extensions, that can then be imported as modules and used transparently in your Python code. This approach may require more work, but can resolve more complex situations. Python's numpy library is an exemplary use of this kind of integration.

The Rust programming language is rapidly becoming a de-facto standard for low level programming: it is a strong contender of C/C++'s place as it offers a similar low-level perspective, but with modern syntax and a safe memory handling paradigm that prevents a lot of common C/C++ bugs and mistakes from happening. PyO3 is a Rust library that provides a way to write Python extensions in Rust; Rust functions can then be called directly from Python. Rust-Cpython is an alternative to PyO3 that supports interoperability between Rust and Python, allowing Rust code to interact with Python objects and call Python functions.

In some cases, it might be sufficient to delegate some computationally heavy tasks to another high level language like Julia, which might still be easier and faster to implement compared to C/C++ or Rust. PyJulia and JuliaCall are two of the most commonly used libraries ta have Python and Julia interact.

For some tasks (like training neural networks, or anything involving large-scale numerical computation and data processing), changing language may not be enough:

it is more efficient to leverage the power of massively parallel computational architectures like modern Graphical Processing Units or dedicated processors. PyCUDA and CuPy are examples of libraries that can be used to interact with NVIDIA GPUs directly from Python. Alternatively, OpenCL provides a framework for writing programs that execute across various types of GPUs and other processors, offering broader hardware compatibility, and can be put to use through the pyopencl library. Additionally, higher-level frameworks like TensorFlow and PyTorch provide transparent GPU acceleration for machine learning and deep learning applications.

By using these integration techniques, developers can optimize the performance-critical sections of their programs without losing the advantages of Python, such as its readability, extensive library support, and active community. This hybrid approach allows for a balanced development process where the strengths of multiple languages are utilized to create efficient and maintainable software solutions.

1.4 Running Python Programs

1.4.1 Scripts and Programs

There is no technical difference between a Python script and a Python program: in theory the terms can be used interchangeably. However, it is common to refer to "scripts" when talking about small and self-contained pieces of code, usually fitting in only one file, used to performa a simple and immediate task and designed to be executed start to finish. They can be run directly by an interpreter and typically do not require user interaction. The term "program" is instead commonly employed to refer to a bigger piece of software, usually structured in multiple files, modules and libraries, with associated tests and likely a broader functional scope. Unlike scripts, programs may involve user interfaces, handle inputs and outputs or maintain state and configuration like a full-fledged software application.

Python provides its own interactive shell, which can be invoked just by launching `python`. However, when experimenting it's well worth trying the extended `ipython` shell, which is a specialized interactive python interpreter that supports useful features like syntax highlighting, auto-completion and command history; dedicated introspection mechanisms can be used to obtain information about objects, functions or modules. This tools are a great aid to the developer, and can save you a lot of time.

1.4.2 Jupyter Notebooks

Jupyter Notebooks are an open-source, locally hosted web application that allows you to create and share documents that contain live code, markdown documentation,

LATEX equations, visualizations, and narrative text. They are widely used in data science, scientific computing, and machine learning due to their interactive nature and the ease with which they facilitate data analysis and visualization. Jupyter Notebooks provide an interactive environment where you can run code in real time. This allows for quick experimentation and immediate feedback, which is particularly useful for data analysis and exploration. Rich Media Support allows to include various types of media such as images, videos, and interactive visualizations directly in the notebook, making it easier to explain and visualize complex concepts. Markdown blocks allow for narrative text to document your workflow and share insights more effectively. All this makes Jupyter Notebooks an excellent tool for teaching and collaborative work. Moreover, notebooks can be easily converted to python scripts or LATEX documentation (for instance, all the examples of this book have been implemented in Jupyter notebooks and exported in LATEX to be included in the corresponding sections). Jupyter Notebooks offer a unique blend of code, visualization, and narrative, making them indispensable for data science, machine learning, and education. They enhance the traditional programming workflow by providing an interactive and collaborative environment, enabling more efficient experimentation, documentation, and sharing of insights.

1.5 Python Kickstart

A complete reference to the Python language can be found in the official documentation at https://docs.python.org/3/.

This introduction assumes the reader to be already familiar with computers in general, and with programming (in any computer language) in particular. Python's syntax and control structures are substantially similar to any other programming language, but there are aspects of the language that is worth being aware of to avoid common mistakes and to quickly become proficient in its use. This introduction covers the basics of the language but also highlights some advanced aspects, which may be useful to both novice and experienced programmers alike.

Python is an interpreted language. A python program consist of one or more source .py files, possibly organized in modules, libraries and packages for convenience of use and distribution. A python program therefore needs an interpreter to be executed; the interpreter (the most common is cpython) takes care of converting the high-level python statements in machine code and executing them.

1.5.1 At the Beginnig

A Python file can optionally start with a "hashbang" comment, which tells the operating system which interpreter to use to run the script contained in the file itself, if the file is executable. The second comment, the encoding line, is useful to declare the encoding of the source file. As the hashbang comment it is also not mandatory, but should always be included if the file uses characters outside of the ASCII range and you expect your script to be executable in a machine diffirent from the one where the script was originally typed in.

```
1   #!/usr/bin/python3
2   # -*- coding: utf8 -*-
```

1.5.2 Scripts, Modules, Libraries, Packages

An isolated python file, intended to be run directly, is usually called a *script*. In general, any python file can be considered a *module* and can be imported by any other python program. Python files intended to be used as modules usually only have code that define members like classes, functions and variables and do not have code outside the scope of functions or classes that would be executed at import time. A directory containing one or more python files is a *package*, for the python interpreter, if it contains a file named __init__.py (which can also be empty). It is often assumed that a *package* is a collection of modules and a *library* is a collection of packages, but most of the times the two terms are used interchangeably.

For example, we could have the file tree:

```
.
|-- module.py
|-- package
|   |-- __init__.py
|   |-- packaged_module_2.py
|   `-- packaged_module.py
`-- script.py
```

where script.py is intended to be run directly, and imports functionality from module.py and from the contents of the package.

External modules and libraries can be included with the import statement:

```
1   import sys
2   info = sys.float_info
```

sys has now become part of the program's namespace and is accessible as any other variable. Importing a whole library is useful, but the resolution of attributes (the . part in sys.float_info) may have have an impact on performances, especially if used in tight loops. Also, it is always a good idea to keep the namespace as small as possible, to avoid potential human mistakes. It is recommended to import only the attributes that are actually needed; the import statement itself also allows to rename the attributes being imported, either to avoid name clashes with existing names or simply to use names which are more appropriate in the current context (for example, a shorter handle for a long name used frequently).

```
ı   from sys import float_info as
    ↪   information_about_type_float
```

In general, the intepreter will search for modules and packages in the directories specified in the list pointed by the variable `path` from the module `sys`. The content of that list can be altered by a script in order to find modules and packages outside of the standard paths.

Beaware that a module is never imported twice. If you are developing interactively and you make changes to a module you have imported, importing the module again is not going to have any effect: use `reload` instead.

1.5.3 Exploring the Language

In Python, all variables are symbolic names that hold a reference to an object. *Everything* is an object (including the basic types likes integers and strings, but also functions!). When an object is assigned to a variable, you can refer to that object by that name, but the data itself is still contained within the object.

For example, the statement:

```
ı   a = 1
```

instantiates an object of type `int` (which happens to refer to the value 1) and creates a reference in the current namespaces named "a", which can be used to refer to such object. `int` happens to be an *immutable* type, which means such an object exists but cannot be modified in any way. This implies that a subsequent statement:

```
ı   a = 2
```

will change the variable a to hold a reference to a new object (2), leaving the other (1) untouched (but eventually collected and destroyed by the garbage collector when its time is due). At the same time, multiple variables can hold reference to the same object. For example:

```
1   l = [1,2,3]
2   k = l
```

creates two references, l and k to an object which in this case is a list of numbers.
Therefore, an alteration of the object is visible by all of its references:

```
1   print('l is: ', l)
2   print('k is: ', k)
3   print('Appending 4 to k')
4   k.append(4)
5   print('l now is: ', l)
6   print('k now is: ', k)
7
```

```
1   l is:  [1, 2, 3]
2   k is:  [1, 2, 3]
3   Appending 4 to k
4   l now is:  [1, 2, 3, 4]
5   k now is:  [1, 2, 3, 4]
```

Referencis another object with one of the two variables leaves of course the other
untouched:

```
1   print('l is: ', l)
2   print('k is: ', k)
3   l = [10,9,8]
4   print('l now is: ', l)
5   print('k now is: ', k)
```

```
1   l is:  [1, 2, 3, 4]
2   k is:  [1, 2, 3, 4]
3   l now is:  [10, 9, 8]
4   k now is:  [1, 2, 3, 4]
```

When a mutable object is passed to a function, modifications to the object inside
the function affect the original object; this can easily lead to unexpected side effects.
When an immutable object is passed to a function instead, modifications inside the
function do not affect the original object.

The Basics: Lists, Tuples, Dictionaries and their Handling Lists are heterogeneous
and mutable:

```
1    l = [] # an empty list
2    l = [6] # a list with one element
3    l = ['a', 1, 3+2j] # an heterogeneous list
4    l = [[1,2],[3,4]] # nested lists
```

Abstracting from details, lists in Python are basically implemented in C as vectors of pointers. Random access is therefore O(1), but each access implies following a pointer to reach the actual object, hence tight loops over (for instance) integers may be slower than one excepts and will most likely be penalized by processor's cache misses. However this is usually not a problem when iterating over complex objects. The list extension operation is also implemented in such a way to have an ammortized constant time cost.

Tuples are essentially immutable lists, hence they behave in a similar way, except that once they are created, they cannot be modified. Keep in mind that, like lists, tuples are also containers of references to objects: the tuple itself cannot be mutated, but the objects they point to can change their state. For instance:

```
1    l1 = [1,2]
2    l2 = l1 # l2 is the same list as l1
3    t1 = (l1, l2) # t1 is a tuple which holds references to
     ↪    l1 and l2
4    t2 = t1 # t2 is the same tuple as t1
5    t2 += (3,) # tuples are immutable, so now t2 is a new
     ↪    tuple
6    l1 += [5] # l1 is mutable, hence l1 and l2 still refer
     ↪    to the same object
7              # (which now has one more element)
```

Python allows for implicit unpackng of tuples. For example:

```
1    a,b,c = (1,2,3) # a <- 1; b <- 2; c <- 3;
2    a,b = b,a # happens _logically_ in parallel: a <- 2, b
     ↪    <- 1
3              # no need to explicitly use a third variable
```

When returning multiple values from a function, always use a tuple (and unpack the result directly!). Using output parameters (passing pointers or references to a function and looking for the results after, hence relying on the side effects of the function call) is not only phylosophically unpythonic, but is a recipe for writing unmaintainable code.

To better understand the implications of immutability, try this little excercise:

```
1   l1 = [1, 2, 3]
2   l2 = l1
3   l2 == l1, l2 is l1 # Both names refer to the same object.
4   l2 += [4, 5]
5   l1, l2
6   l1 == l2, l2 is l1 # Mutations of the object are visible
    ↪   through both names.
7
8   t1 = (1, 2, 3)
9   t2 = t1
10  t2 == t1, t2 is t1 # Again, both names refer to the same
    ↪   object.
11  t2 += (4, 5) # But tuples are immutable. The original
    ↪   object cannot be modified, so a new object is
    ↪   created and bound to t2. The original remains
    ↪   unaltered.
12  t1, t2
13  t1 == t2, t2 is t1 # t1 continues to refer to the
    ↪   original object.
14
15  i1 = 0
16  i2 = i1
17  i2 == i1, i2 is i1 # Again, only one object ...
18  i2 += 1 # ... which is immutable, so a new one is
    ↪   created and bound to i2, .
19  i1, i2
20  i1 == i2, i2 is i1; # i1 continues to refer to the
    ↪   original object.
```

Strings are immutable collections of unicode characters. There is no separate "character" type, indexing a string produces strings of length 1. That is, for a non-empty string s, `s[0] == s[0:1]`.

```
1   a = 'a string'
2   b = "another string"
3   c = """A multi-line string!
4   In here you can use a single
5   " or ' without escaping it.
6   """
7   d = ("String literals "
8       "that are part " "of a single expression"
9       "and have only whitespace between them "
10      "will be implicitly converted to a single string
        ↪   literal")
11  e = "A string with a ' in the middle"
12  f = 'the \' can also just be explicitly escaped'
13  g = "Same for the \""
```

Lists (and tuples) elements can be accessed directly by index:

```
1   a = [11,22,33,44,55,66,77,88,99,110]
2   a[3]    # is 44, indexing is 0 based!
3   a[-3];  # is 88, negative index is offset from the end
    ↪   of the sequence
```

Sublists can be produced by slicing. Elements can be contiguos:

```
1   a[3:6] # is the slice [44,55,66]. Lower bound inclusive,
    ↪   upper exclusive
2   a[8:-2]# is the slice [88] (the right bound is
    ↪   exclusive!)
3   a[4:3] # is empty!
4   a[:3]   # leaving out a limit implicitly assumes to reach
    ↪   the end.
5           # this is equivalent to a[0:3]
6   a[5:];  # and this is equivalent to a[5:10] in this
    ↪   particular case
```

or selected every n-th:

```
1   a[0:8:2] # is [11,33,55,77]: selects every 'step size'
    ↪   element
2   a[0:8:3]; # is therefore [11,44,77]
```

Slicing can also be used to write elements and sublists:

```
1   a[1:-1] = 'x' # a becomes [11,'x',110]
2   a[-2:] = 'yzk' # [11,'x',110] becomes [11, 'y', 'z', 'k']
3   a[:1] = ['i', 'x'] # [11, 'y', 'z', 'k'] -> ['i', 'x',
    ↪   'y', 'z', 'k']
```

Lists can be extendend by summing lists or appending elements:

```
1   a = list(range(0,4)) # [0,1,2,3]
2                         # range() is a type
3   a += [5,6] # [0,1,2,3,5,6]
4   a.append(7) # [0,1,2,3,5,6,7]
```

Lists can also be extended by multiplication:

```
1   b = [1,2]
2   b *= 3 # [1,2,1,2,1,2]
```

For instance, let's instantiate a matrix and set one of its elements:

```
1   c = [[0]*3]*3 # [[0, 0, 0], [0, 0, 0], [0, 0, 0]]
2   c[0][0] = 1
3   c[0][1] = 2
4   c[2][2] = 3
```

What are the contents of the matrix?

```
1   print(c)
```

```
1   [[1, 2, 3], [1, 2, 3], [1, 2, 3]]
```

Did this surprise you? [0] is a list that contains only 0, [0]*3 is a list of thre immutable objects. But [0,0,0]*3 is a list of three references to the the same mutable object!

The in operator allows to check for the presence of elements in any iterable:

```
1   3 in a; # True, 3 is an element of the list a
2   3 in c; # False, c is a list of lists!
```

in particular, in a list will be a linear search, but other iterable types, such as sets and dictionaries, have specific implementations that make the search orders of magnitude faster, and in some cases $O(1)$.

Dictionaries are an efficient and effective way of representing and handling key-value data. In Python, any *hashable* object can be used as a key:

```
1   d = {
2       'first key': 'first value',
3       2: 'second value',
4       'yet another': 0.1234,
5       'print me': print,
6        print: 'function print!',
7   }
8
9   print(d['first key'])
```

```
| first value
```

Yes, any object can potentially be referenced in a dict, even functions!

```
| d['print me']('something to print')
```

```
| something to print
```

```
| print(d[print])
```

```
| function print!
```

Elements can also be removed from dictionary using the `del` keyword:

```
1  print('before:', d)
2  del d['first key']
3  print('after:', d)
```

```
| before: {'first key': 'first value', 2: 'second value',
   ↪ 'yet another': 0.1234, 'print me': <built-in
   ↪ function print>, <built-in function print>:
   ↪ 'function print!'}
2  after: {2: 'second value', 'yet another': 0.1234, 'print
   ↪ me': <built-in function print>, <built-in function
   ↪ print>: 'function print!'}
```

Keep in mind that dictionaries are *unordered*. They are visually presented in some order, and they can be iterated over, but the order is meaningless and bound to change when the dictionary changes in size. If the order is important, use lists or tuples or have a look a the `OrderedDict` in `collections` module.

Variables, Binding, Call-by-Value In statically typed languages, such as C, types are associated to varibales: variables are to be explicitly declared with the type they are holding a reference or value to. If you try to to assign something different to it, the compiler will complain. In python, instead, variables do not have types: objects do. Variables don't neet do be declared, as they are never assigned a value: they are *bound* to an object. For example:

```
1    i = 3
2    i = 'hello'
```

here 'i' is bound to an object of type int, and subsequently to an object of type string. 'i' doesn't care what type object it's bound to, as the type checks will be performed at runtime against the object itself.

Python uses *call-by-value* semantics, as if a called function would receive a copy of the argument's value. In reality, however, all values (variable) are references, and what is passed to the function is a copy of the *reference* to the actual object itself. Changes made to the object inside the function will affect the original object outside the function (unless the object itself is immutable).

First-Class Objects Almost everything in Python is a first-class object (including functions, classes, modules). This means that they can be bound to variables, used in data structures and passed to functions just like any other value. Note that it is normal in python to refer to a function without wanting to call it (to bound it to some variable or in a data structure and call it at some later time).

Boolean Contexts In general, in Python it is considered:

- False

  - Numbers equal to zero: `0`, `0.0`, `0j`
  - Empty sequences: [], {}, "", ()
  - `None` is an object that denotes the absence of an object
  - `False`

- True

  - Pretty much everything which is not evaluating to `False`
  - `True`

- User definet types can determine their own interpretation in boolean context by implementing `__bool__`.

Indentation, Blocks and Flow Control Python uses only indentation as delimitation mechanism for blocks. This may look strange to programmers coming from explicitly delibitate blocks like C or java, but offers some advantages:

- removes visual clutter caused by the presence of block delimiters;
- avoids wasting lines on block delimitation tokens, which in turns makes the code shorter and hence easier to read;
- ensure that the block structure perceived by the compiler is the same as that perceived by the programmer. In languages like C or java, it is possible to format the code in such a way that implies the presence (or the absence) of a block, but to have a (misplaced) parethesis change the actualy meaning of the code for the compiler in such a way that is potentially difficult for the human eye to see at first glance.

Below is a straightforward example of an `if` – `else` statement; notice how the indentation naturally delineates the blocks scopes:

```python
a = 1
b = True
c = 'some string'

if (a and b):
    l = []
    d = {}
elif (b or c):
    d = {}
    if c:
        l = []
else:
    t = ()
```

Flow can also be controlled with a switch statement, which in Python is called `match`:

```python
match a:
    case 1:
        pass
    case 2 | 3:
        pass
```

The `pass` keyword allows to have an 'empty' block. Loops in Python are also natural. In general, any sequence can be iterated over:

```python
for c in 'hello':
    print(c)
```

```
h
e
l
l
o
```

or using a condition:

```python
eps = 1.0
iters = 0
while 1+eps > 1:
```

```
4          iters += 1
5          eps /= 2
6     print(iters)
```

```
1     53
```

in both cases, the cycle can be interrupted with `break`, which immediately terminates the loop, or the rest of the current iteration can be skipped with `continue`, which goes on directly with the next iteration. Both `for` and `while` admit an optional `else` clause that is executed onl if the loop runs to completion, without being cut short by `break`, `return` or `raise`:

```
1     for i in [1,2,3,4,5,6]:
2         if i % 2 == 0:
3             continue
4
5         if i > 4:
6             break
7     else:
8         print('The cycle was not broken')
```

Functions Functions are defined with the `def` statement:

```
1     def do_something(arg1, arg2):
2         """This docstring documents the function"""
3         # Some implementation
4         return a, b, c
```

Multiple return values are implicitly packed into a tuple, and can be unpacked directly when calling the function if necessary:

```
1     val1, val2 = 1, 2
2     retval1, retval2, retval3 = do_something(val1, val2)
```

If a function does not have a return statement, it implicitly returns `None`.

Lexical Scopes Python resolves names following the L^{n}GB rule: Local (nested), Global, Builtins. In particular, an unqualified (missing an explicit scope) name is looked up in the following order:

- L: local: inside a function body, look for a binding in the function's local scope. Outside of a function, jump to B
- n: nested: progress outwards through any sorrounding nested functions, looking in the respective local scopes (the closure example we made before applies exactly this rule to find n)
- G: global: search in the global scope (variable that have been explicitly declared globals, or defined at the module level)
- B: builtins: look for a binding in the builtins
- If no binding was found, raise a `NameError`.

If a name is bound inside a function (as a parameter or with an explicit assignment), the binding is local to that function. Variables outside the local scope can be assigned after being referred explicitly using the `global` keyword:

```
1   a, b = 0, 1
2   def f(a):
3       global b
4       a = 2
5       b = 3
6   f(a)
7   print('a, b are now: ', a, b )
```

```
1   a, b are now:  0 3
```

Lexical Closures Consider the following example:

```
1   def make_adder(n):
2       def adder(x):
3           return n+x
4       return adder
```

Here the function `make_adder` returns the function `adder` which refers to a name (n) which is bound to the scope of the outer function. In particular:

- When the outer function returns, n goes out of scope, and would normally disappear along with its binding;
- However, the inner function `adder` still references it, so is in fact prolonging its lifetime to be able to use it when it will be called;
- Each call to the outer `make_adder` function creates and returns a new, different version of `adder`, with its own distint binding of n.
- The inner binding of n is called a *closure* of n.

- Notice that such a closure is a combination of function and state, wich is the basis of object oriented programming. The returned `adder` function is in fact a first-class object which retain parts of its environment as state.

Passing Arguments In python, functions can be defined with any number of positional arguments and keyword arguments. Positional arguments are identified by their position in the function signature, and are always required when calling a function. Keyword arguments (also called parameters) are instead defined by their name, and are optional with a default value defined in the function's signature.

```
1   def f(arg1, arg2, arg3, kwarg1=False, kwarg2=True,
    ↪   kwarg3=list()):
2       pass
```

Functions can also be defined with a variable number of arguments, using the special syntax `*` and `**`as in:

```
1   def f(*arguments_list, **kword_args_dict):
2       print(arguments_list, kword_args_dict)
3
4   f( 1, 2, 3, x=4, y=5)
```

```
1   (1, 2, 3) {'x': 4, 'y': 5}
```

in this way, `f` can be called with any number of arguments and keyword parameters, which get assembled in a tuple bound to the variable `arguments` and a dictionary `kword_args`. Arguments and parameters can then be unrolled with the same syntax:

```
1   def real_f(a, b, c=False, d=True):
2       return a, b, c, d
3
4   def proxy(*args, **kwargs):
5       if 'd' not in kwargs:
6           kwargs['d'] = 'This was missing'
7       return real_f(*args, **kwargs)
8
9   print(real_f(1,2,c='changed'))
10  print(proxy(1,2,c='changed'))
```

```
1   (1, 2, 'changed', True)
2   (1, 2, 'changed', 'This was missing')
```

λ-Functions In Python it is possible (and oftentimes convenient) to use anonymous (not associated with a name) functions. Those are called lambda-functions and can be defined on the fly:

```
1   lambda a,b,c: a*b+c
```

```
1   <function __main__.<lambda>(a, b, c)>
```

In general, lambda functions are useful in context where they have to be used only once, and they are not significant in the scope of the code. For example, a function that computes a key to use to sort a list, or some simple operation to be applied to every element of a list, etc.:

```
1   x = (lambda a,b,c: a*b+c)(1,2,3)
2   x = map( lambda a: a+1, [1,2,3,4])
3   x = filter( lambda x: x%2 == 0, [1,2,3,4])
4
5   from functools import reduce
6   x = reduce( lambda a,b : a+b, [1,2,3,4])
```

Notice how all of the above could also be implemented explicitly using a loop, but this compact syntax is usually less error prone and much easier to read.

Dealing with Exceptions Exceptions are a simple way of dealing with errors, or more in general, with situations that do not belong to the linear flow of a piece of code. They can be thought of as a way of signaling that some condition occurred, and specifying what can or should be done about it. Exception can be raised with the rise keyword, and can be handled within an try except block:

```
1   try:
2       print('Rising exception')
3       raise Exception('Something exceptional!')
4   except Exception as e:
5       print('Handling exception:', e)
```

```
1  Rising exception
2  Handling exception: Something exceptional!
```

Different exceptions can be handled depending on their type, be re-rised as-is to be handled by the caller, and an optional `finally` block allows for code that is always executed regardless of wether an exception occured.

```
1  try:
2      print(dict()['nonexistent key'])
3  except IndexError:
4      print('It was an IndexError')
5      raise
6  except KeyError:
7      print('It was a KeyError')
8      raise
9  finally:
10     print('This is always executed')
```

```
1  It was a KeyError
2  This is always executed
```

```
1
2              ------------------------------------------------┘
   ↪   -------------------------
3
4          KeyError
   ↪   Traceback (most recent call last)
5
6          Cell In[43], line 2
7            1 try:
8  ----> 2      print(dict()['nonexistent key'])
9            3 except IndexError:
10           4     print('It was an IndexError')
11
12
13         KeyError: 'nonexistent key'
14
```

There are a plethora of exceptions types that can be imported and used directly from the `builtins` module; otherwise you can define any exception with extended behaviour by extending `Exception` with a new class. Exceptions are naturally organized in an inheritance hierarchy; catching an exception will also catch all its subtypes.

```
1  class my_exception(Exception):
2      pass
3  class subtype_exception(my_exception):
4      pass
5
6  try:
7      raise subtype_exception()
8  except my_exception:
9      print('cought you!')
10
```

Asking for Permission versus Asking for Forgiveness Defensive programmers tend to prevent error situations by "asking for permission" first (also referred to as "look before you leap"), while in Python the preferred way is to "ask for forgiveness" (also referred to as handling exceptions). For example, compare the two approaches in the following code:

```
1  def divide_AfP(num, den):
2      if den != 0:
3          return num/den
4      else:
5          return 0
6
7  def divide_AfF(num, den):
8      try:
9          return num/den
10     except ZeroDivisionError:
11         return 0
```

The latter function is the better approach in Python. It is usually faster: there is no branching involved if the division can be done. The first approach would potentially be faster only in case the probability of having den = 0 would be the same as den != 0. Also, the AfF approach is usually more readable, as the intention of the code is not clouded by a number of checks that may hide its purpose. It is also usually easier to add an additional exception (and relative recovery code) that to devise conditions to check first. Moreover, preemptive checks may have a computational cost that in the AfF case is completely avoided in the (hopefully) vastly more frequent case in which an exception does not arise.

```
1
2  processed_list = [process(el) for el in iterable]
3
4
5  try:
6      process(d)
```

```
7   except Exception as e:
8       raise e
9
10  processed_generator = (process(el) for el in iterable)
11
```

Classes and Objects Classes bundle data and functionality together. A class defines a *type* of object, and allows *instances* of that type to be made. A basic class definition looks like this:

```
1   class MyClass:
2
3       def __init__(self, init_value_1):
4           self.counter = init_value_1
5           self.__myclassonly = True
6
7       def increment(self):
8           self.counter += 1
9           return self.counter
10
11      def substract(self, amount):
12          self.counter -= amount
13          return self.counter
14
```

MyClass is a class that defines a new type. increment and substract are ordinary methods. __init__ is a *magic* method called a "constructor", which is automatically invoked when creating a new object and performs all the required initialization. Inheritance allows to extend or alter the behavior of it's base class:

```
1   class MyExtendedClass(MyClassName):
2
3       def __init__(self, init_value_1, init_value_2):
4           super(MyExtendedClass,
5            ↪  self).__init__(init_value_1)
            self._v2 = init_value_2
6
7       def add(self, amount):
8           self.counter += amount
9           return self.counter
10
11      def increment(self):
12          self.v2 += 1
13          return super(MyExtendedClass, self).increment()
```

MyExtendedClass has an extra attribute, and an extra method compared to MyClass. Also, the method increment has been redefined to do something extra with respect to the inherited method. Multiple inheritance is supported: a class can provide a comma sperated list of classes and the method resolution order (MRO) will decide which is the correct method to use depending on the order of inheritance (or raise an error if that's not possible to determine). You can also explicitly choose to call a specific class method from the subclass by naming the class itself. Here is an example:

```python
class A():
    name = 'classA '
    def test(self):
        print('testA' + self.name, end=' ')

class B():
    name = 'classB '
    def test(self):
        print('testB' + self.name, end=' ')

class C(A,B):
    name = 'classC '
    def test(self):
        print('testC' + self.name, end=' ')
        super(C, self).test()

class D(B,A):
    name = 'classD '
    def test(self):
        print('testD' + self.name, end=' ')
        super(D, self).test()

class E(A,B):
    name = 'classE '
    def test(self):
        print('testE' + self.name, end=' ')
        A.test(self)
        B.test(self)

C().test()
print('\n--')
D().test()
print('\n--')
E().test()
```

```
testCclassC  testAclassC
--
testDclassD  testBclassD
--
testEclassE  testAclassE  testBclassE
```

Python does not enforce privacy by restricting access to attributes and methods. However, by convention, you should not access names that start with an underscore (like _v2), which are to be considered private. Names starting with two underscores (like __myclassonly) are instead not meant to be used by sub classes, and are mangled by the interpreter; this is intendded to be a protection against name clashes with subclasses. Names starting and ending with double underscore (like __init__) are special names.

In statically typed languages, polymorphism is usually tightly coupled to inheritance or interfaces (like in Java). This means that an object can only be swapped at runtime with another object which implements the same interface, or that subclasses the same class. In python instead polymorphism does not need inheritance, because it implements *duck typing*: "if it looks like a duck, and it quacks like a duck, then it is a duck!". With this approach, what an object can do is more important than it's type. The approach is to `try` to access an attribute or call a method: if the object does implement that, we're good. If it doesn't, we can then `catch` the exception and decide how to recover from the situation. This allow your code to handle objects of any type as long as it has the methods and attributes you need.

1.6   Closing Remarks

This chapter is of course just an introduction that barely skims the surface of many really important topics. It is intended to give an overview of what is the state of the art, and remark the most important concepts to get started. Regarding "the art" of software development, [7–10] are great starting point. References [11–14] are also great (and fun!) resources which bring lots and lots of invaluable experience to the table. References [15–17] are excellent resources to get started with project management and version control. One of the best ways to delve into python is of course through the official documentation at https://docs.python.org/3/. and extensive use of `ipython` to experiment. Eckel [18] is not about Python but another object oriented language (Java), however it is a strongry recommended reading as it provides an amazingly clear explaination of object oriented programming, type systems, interfaces and so on which apply to any language. Design patterns can be of great help in complex projects [4–6].

Chapter 2
Essential Mathematics Tools

This chapter introduces the fundamental mathematical tools used throughout the book. The focus is on simplicity, prioritizing real-life applications over formal definitions to clarify critical mathematical concepts. The topics covered are essential for solving systems of ordinary differential equations and understanding their real-world connections. Some topics touch on advanced mathematical concepts, many of which are still at the forefront of research and would require entire books to grasp fully. However, users of mathematical models-who are often not mathematicians-must have an intuitive sense of the complexities that may lie behind numerical solvers and the critical aspects that require careful attention. Developing effective models based on reality is rarely straightforward, and complexity can quickly outstrip human intuition. The behavior of chaotic systems, unstable methods, ill-conditioned operations, and systematic errors can lead to unexpected outcomes. A basic understanding of what happens "behind the scenes" and the ability to identify potential issues when things go wrong are crucial for maintaining accuracy and correctly interpreting the results of any computational study. Python will frequently used to solve the mathematical problems discussed in this chapter.

2.1 Linear Algebra (with Python)

To accurately replicate a physical phenomenon in a computer simulation, it is crucial to incorporate relevant data associated with the phenomenon under investigation. For instance, when modeling specific brain regions, gather data on neuron activity or the concentrations of substances like neuromodulators. Managing these extensive datasets and performing operations on them can often be complex. Linear algebra offers the mathematical tools to streamline and efficiently process these datasets. This section explains how these tools aid in succinctly formulating mathematical equations that describe a given phenomenon. The focus will be on primary characteristics

© The Author(s), under exclusive license to Springer Nature Singapore Pte Ltd. 2025
D. Caligiore and S. Carli, *Simulating the Brain*, Brain Informatics and Health,
https://doi.org/10.1007/978-981-96-2718-9_2

and operations related to three fundamental mathematical objects in linear algebra: scalars, vectors, and matrices. Formal aspects such as theorem proofs are intentionally not addressed, favoring an intuitive approach grounded in Python. References will be available for readers interested in the formal mathematical elements.

2.1.1 Scalars, Vectors, Matrices

In linear algebra, the three fundamental entities are scalars, vectors, and matrices. Scalars are individual numbers and typically belong to a field, such as the real or complex numbers. They are used to scale vectors or matrices during operations (Fig. 2.1).

Vectors are one-dimensional ordered arrays of numbers that represent both direction and magnitude in space. They can be displayed as rows, columns, or ordered sets of numbers. For instance, a vector comprising three elements may be represented as follows: $\mathbf{v} = (2, 3, 1)$ (Fig. 2.2). Vectors are identified by a single index that refers to specific elements within them. For example, in the vector $\mathbf{v}$, v_2 corresponds to the second value, which is 3. By implementing a Cartesian reference system, we can map the plane onto the set $\mathbb{R}^2$ comprising ordered pairs of real numbers. Each point on this plane can be paired with a vector, conceptualized as an oriented segment extending from the origin to the point itself. Consequently, we can equate the vector $\overrightarrow{OP}$ with the ordered pair (x, y), representing the coordinates of point P. Thus, instead of denoting $\overrightarrow{V}$ as $\overrightarrow{OP}$, we can simply write $\overrightarrow{V} = (x, y)$, where x and y are the scale components of the vector. To compute the norm (or magnitude) of the vector $\overrightarrow{V} = (x, y)$, we can use the formula for Euclidean distance. In particular, the norm of a vector in 2D space is

Fig. 2.1 Vectors $\vec{v}$ and $c\vec{v}$ in the Cartesian plane with grid lines

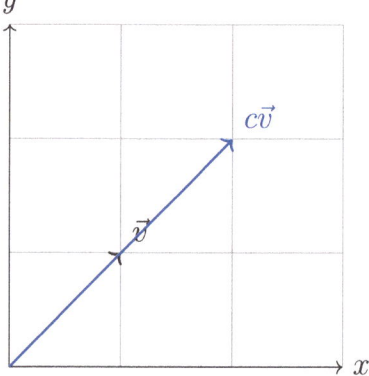

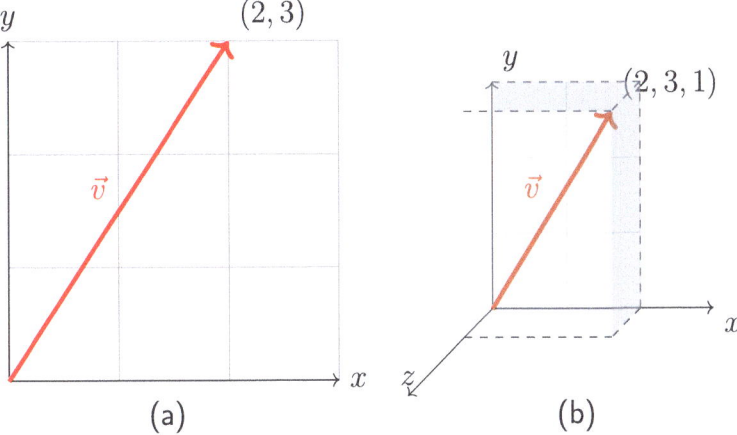

Fig. 2.2 **a** A vector $\vec{v}$ in the Cartesian plane. **b** A vector $\vec{v}$ in the Cartesian space $\mathbb{R}^3$

$$\|\overrightarrow{V}\| = \sqrt{x^2 + y^2}.$$

The discussion made for a vector in $\mathbb{R}^2$ can be extended to $\mathbb{R}^3$ and, in general, to the n-dimensional space $\mathbb{R}^n$. Similarly, in $\mathbb{R}^3$, the norm of a vector $\mathbf{v} = (v_1, v_2, v_3)$ is computed as $|\mathbf{v}| = \sqrt{v_1^2 + v_2^2 + v_3^2}$. For the n-dimensional space $\mathbb{R}^n$, the norm of a vector $\mathbf{v} = (v_1, v_2, \ldots, v_n)$ is given by $|\mathbf{v}| = \sqrt{v_1^2 + v_2^2 + \cdots + v_n^2}$.

Matrices represent structured arrays of numbers organized into rows and columns, offering essential tools for depicting linear transformations, systems of linear equations, and various mathematical concepts. Comprised of individual scalars, matrices facilitate fundamental operations such as addition, subtraction, and multiplication.

A matrix consists of a structured 2D array of numbers with two indices: one indicating the row and the other the column. They play a pivotal role in various mathematical contexts, offering diverse shapes and properties, each serving distinct purposes in linear algebra and beyond. Below are some examples of different kinds of matrices.

$$A = \begin{bmatrix} 1 & 2 & 3 \\ 4 & 5 & 6 \\ 7 & 8 & 9 \end{bmatrix} \qquad A^T = \begin{bmatrix} 1 & 4 & 7 \\ 2 & 5 & 8 \\ 3 & 6 & 9 \end{bmatrix} \qquad B = \begin{bmatrix} 1 & 2 & 3 \\ 4 & 5 & 6 \end{bmatrix} \qquad C = \begin{bmatrix} 1 & 2 & 3 \\ 2 & 4 & 5 \\ 3 & 5 & 6 \end{bmatrix}$$

(a) \qquad\qquad\qquad (b) \qquad\qquad\qquad (c) \qquad\qquad\qquad (d)

$$D = \begin{bmatrix} 0 & -2 & 3 \\ 2 & 0 & -5 \\ -3 & 5 & 0 \end{bmatrix} \qquad E = \begin{bmatrix} 2 & 0 & 0 \\ 0 & 4 & 0 \\ 0 & 0 & 6 \end{bmatrix} \qquad F = \begin{bmatrix} 1 & 0 & 0 \\ 0 & 1 & 0 \\ 0 & 0 & 1 \end{bmatrix}$$

$$\text{(f)} \hspace{8em} \text{(g)}$$

$$\text{(e)}$$

Examples of different kinds of matrices: (a) a square matrix, it has an equal number of rows and columns; (b) the transpose matrix of the matrix A, it swaps its rows with its columns; (c) a rectangular matrix, it has different numbers of rows and columns; (d) a symmetric matrix, it is equal to its transpose; (e) a skew-symmetric matrix, it is equal to the negative of its transpose; (f) a diagonal matrix, it has non-zero elements only on its diagonal; (g) an identity matrix, it has ones on its main diagonal and zeros elsewhere.

These examples illustrate the diversity of matrices, showcasing squared, rectangular, symmetric, diagonal, and other types, each with unique properties and applications in mathematical analyses and computations.

In Python, there exist numerous techniques for creating and initializing vectors and matrices. Let's delve into the primary ones, focusing on the utilization of the NumPy library.

```python
# Importing the NumPy library
import numpy as np

# Create a vector v of three elements using NumPy array
v = np.array([2,3,1])

# Create a vector v of n integers numbers using NumPy
#    arange()
n = 10
v = np.arange(n)
```

The default behavior of arange() generates integers, but it allows for specifying the data type as well. In addition, it is also possible to generate numbers with a step.

```python
# Create a vector of n floats numbers using NumPy
#    arange()
v = np.arange(n, dtype='float')

# Using arange() to create a vector of numbers from 1 to
#    9 with a step of 2 as integers
v = np.arange(1, 10, 2)
```

The NumPy method arange() and the built-in Python function range() both produce sequences of numbers, yet they exhibit critical distinctions. The arange() method yields a NumPy array that can accommodate elements of any data type, primarily used for generating numerical sequences tailored for array computations. Conversely, the range() function generates a range object that delivers integers, commonly utilized for iterating through sequences of numbers within loops. Below is a straightforward code example demonstrating this disparity. Additionally, we illustrate how to initialize a vector using range().

```
1   # Initializing a vector using range function
2   v = list(range(1, 11))  # Creates a vector containing
    ↪   numbers from 1 to 10
```

In this example, range(1, 11) generates numbers from 1 to 10 (inclusive of 1 but exclusive of 11), and then list(range(1, 11)) converts these numbers into a list, resulting in the vector [1, 2, 3, 4, 5, 6, 7, 8, 9, 10]. Below is a brief overview of other methods for initializing a vector.

```
1   # Creating a vector v using NumPy linspace() function
2   # The linspace function generates an array of 100 evenly
    ↪   spaced numbers between 1 and 10 (inclusive)
3   v = np.linspace(1, 10, 100)
4
5   # Creating a vector w using NumPy ones() function
6   # The ones() function generates an array of size 10 with
    ↪   all elements set to 1
7   w = np.ones(10)
8
9   # Creating a vector z using NumPy zeros() function
10  # The zeros() function generates an array of size 10
    ↪   with all elements set to 0
11  z = np.zeros(10)
12
13  # Creating a vector y using NumPy random.rand() function
14  # The random.rand() function generates an array of size
    ↪   10 with elements sampled from a uniform distribution
    ↪   over [0, 1). The  the interval [0, 1) includes all
    ↪   real numbers greater than or equal to 0, up to but
    ↪   not including 1.
15  y = np.random.rand(10)
```

When we use the random.rand() function it is possible to set the seed for the random number generator to a given number, ensuring reproducibility of results. Below an example using 20 as seed.

```
1   # Setting the seed for reproducibility
2   # When you set the seed using np.random.seed(), it
    ↪   initializes the random number generator
3   # This ensures that the sequence of random numbers
    ↪   generated will be the same every time you run the
    ↪   code
4   np.random.seed(seed=20)
5
6   # Creating a vector y containing 10 random numbers
    ↪   between 0 and 1
7   # The numbers are sampled from a uniform distribution
    ↪   over the interval [0, 1)
8   y = np.random.rand(10)
```

The functions utilized for creating and initializing a vector can also be employed for creating and initializing a matrix.

```
1   # Import NumPy library to utilize its functionalities
2   import NumPy as np
3
4   # Create a 2x3 matrix using NumPy array function
5   m = np.array([[1, 2, 3], [4, 5, 6]])
6
7   # Create a 2x3 matrix filled with ones using np.ones()
    ↪   function
8   m = np.ones((2, 3))
9
10  # Create a 2x3 matrix filled with zeros using np.zeros()
    ↪   function
11  m = np.zeros((2, 3))
12
13  # Set the random seed to ensure reproducibility of
    ↪   results
14  np.random.seed(20)
15
16  # Generate a 2x3 matrix with random values from a
    ↪   uniform distribution using np.random.rand() function
17  m = np.random.rand(2, 3)
```

With NumPy, we can effortlessly create various types of matrices.

```
1    # Define a 2x2 matrix
2    m = np.array([[1, 2], [10, 20]])
3
4    # Transpose the matrix using np.transpose() function
5    mT = np.transpose(m)
6
7    # Create a 3x3 identity matrix using np.identity()
     ↪    function
8    m = np.identity(3)
9
10   # Create a 3x3 matrix with random values on the diagonal
     ↪    and zeros elsewhere
11   m = np.eye(3, dtype=int) * np.random.rand(3)
12
13   # Generate a 3x3 matrix with random values
14   m = np.random.rand(3, 3)
15
16   # Extract the diagonal elements of the matrix using
     ↪    np.diag() function
17   d = np.diag(m)
18
19   # Generate a 3x3 matrix with random values
20   a = np.random.rand(3, 3)
21
22   # Extract the lower triangle of the matrix
23   lower_triangle = np.tril(a)
24
25   # Transpose the lower triangle of the matrix and zero
     ↪    elements above the main diagonal
26   transposed_lower_triangle = np.tril(a, -1).T
27
28   # Construct a symmetric matrix by adding the lower
     ↪    triangle and its transposed form
29   m = lower_triangle + transposed_lower_triangle
```

2.1.2 Computational Rules

Linear Algebra operates on a set of computational rules that govern the manipulation and transformation of vectors and matrices. In this paragraph, we explore fundamental computational rules in linear algebra, including the product of a scalar for a vector, the sum of vectors, product among vectors, scalar product.

2.1.2.1 Product of a Scalar for a Vector

One of the fundamental operations in linear algebra is the scalar-vector multiplication. This operation involves multiplying each component of a vector by a scalar value. Mathematically, if $\vec{v}$ is a vector and c is a scalar, then the product of c and $\vec{v}$, denoted as $c\vec{v}$, is a new vector whose components are obtained by multiplying each component of $\vec{v}$ by c.

$$c\vec{v} = \begin{bmatrix} c \cdot v_1 \\ c \cdot v_2 \\ \vdots \\ c \cdot v_n \end{bmatrix}$$

Here, $v_1, v_2, \ldots, v_n$ are the components of vector $\vec{v}$.

Let's consider a simple example in two dimensions. Suppose we have a vector $\vec{v} = \begin{bmatrix} 1 \\ 1 \end{bmatrix}$ and a scalar $c = 2$. To visualize the scalar-vector multiplication, we can represent $\vec{v}$ as an arrow in the Cartesian plane, starting from the origin and ending at the point $(1, 1)$. Multiplying $\vec{v}$ by c results in a new vector $c\vec{v}$ with components $(2 \times 1, 2 \times 1) = (2, 2)$. We can represent $c\vec{v}$ as an arrow starting from the origin and ending at the point $(2, 2)$ (Fig. 2.1).

2.1.2.2   Sum of Vectors

Another essential operation in linear algebra is the addition of vectors. When adding two vectors, corresponding components are added together to obtain the components of the resulting vector.

Mathematically, if $\vec{v}$ and $\vec{w}$ are two vectors, then their sum $\vec{v} + \vec{w}$ is a new vector whose components are obtained by adding the corresponding components of $\vec{v}$ and $\vec{w}$.

$$\vec{v} + \vec{w} = \begin{bmatrix} v_1 + w_1 \\ v_2 + w_2 \\ \vdots \\ v_n + w_n \end{bmatrix}$$

Consider two vectors $\vec{v} = \begin{bmatrix} 2 \\ 3 \end{bmatrix}$ and $\vec{w} = \begin{bmatrix} 6 \\ 1 \end{bmatrix}$. To visualize the sum of these vectors, we represent $\vec{v}$ and $\vec{w}$ as arrows in the Cartesian plane. The resultant vector $\vec{v} + \vec{w}$ is obtained by adding the corresponding components of $\vec{v}$ and $\vec{w}$:

$$\vec{v} + \vec{w} = \begin{bmatrix} 2 + 6 \\ 3 + 1 \end{bmatrix} = \begin{bmatrix} 8 \\ 4 \end{bmatrix}.$$

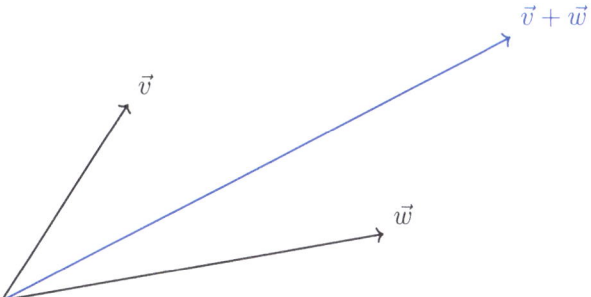

In this representation, the sum vector $\vec{v} + \vec{w}$ illustrates how vector addition results in a new vector pointing from the origin to the coordinates $(8, 4)$ in the Cartesian plane.

2.1.2.3 Product Among Vectors

The product among vectors encompasses various operations, including the dot product, the cross product, and the element-wise (Hadamard) product, each serving distinct purposes in linear algebra and geometry. The dot product yields a scalar value, the cross product results in a vector perpendicular to the plane containing the original vectors, and the Hadamard product is an element-wise product resulting in another vector of the same dimension.

Scalar Product. The scalar product, also known as the dot product, is a fundamental operation in linear algebra that yields a scalar quantity. It is defined as the sum of the products of corresponding components of two vectors.

Mathematically, if $\vec{v}$ and $\vec{w}$ are two vectors, then their scalar product, denoted as $\vec{v} \cdot \vec{w}$, is given by:

$$\vec{v} \cdot \vec{w} = v_1 \cdot w_1 + v_2 \cdot w_2 + \cdots + v_n \cdot w_n$$

Here, $v_1, v_2, \ldots, v_n$ and $w_1, w_2, \ldots, w_n$ are the components of vectors $\vec{v}$ and $\vec{w}$, respectively.

Cross Product. The cross product is a vector operation that produces another vector, perpendicular to both original vectors. This operation is only defined for vectors in three-dimensional space.

If $\vec{v}$ and $\vec{w}$ are two 3D vectors, their cross product, denoted as $\vec{v} \times \vec{w}$, is defined as:

$$\vec{v} \times \vec{w} = \begin{vmatrix} \hat{i} & \hat{j} & \hat{k} \\ v_1 & v_2 & v_3 \\ w_1 & w_2 & w_3 \end{vmatrix}$$

This results in a new vector:

$$\vec{v} \times \vec{w} = (v_2 w_3 - v_3 w_2)\hat{i} + (v_3 w_1 - v_1 w_3)\hat{j} + (v_1 w_2 - v_2 w_1)\hat{k}$$

The cross product has applications in physics, particularly in calculating torque, angular momentum, and electromagnetic force.

Element-wise (Hadamard) Product. The element-wise or Hadamard product is another vector operation where each component of one vector is multiplied by the corresponding component of the other. Unlike the dot product, the Hadamard product results in a vector rather than a scalar.

If $\vec{v}$ and $\vec{w}$ are two vectors of the same dimension, their Hadamard product, denoted as $\vec{v} \circ \vec{w}$, is defined as:

$$\vec{v} \circ \vec{w} = \begin{bmatrix} v_1 \cdot w_1 \\ v_2 \cdot w_2 \\ \vdots \\ v_n \cdot w_n \end{bmatrix}$$

The Hadamard product is useful in various computational applications, particularly when manipulating vectors in an element-wise fashion, as commonly seen in machine learning and data analysis tasks. These computational rules form the foundation of many linear algebra operations, enabling the manipulation and analysis of vectors and matrices in diverse mathematical and computational contexts.

2.1.2.4 Matrix Product

The product of two matrices involves multiplying each element of a row from the first matrix with the corresponding element of a column from the second matrix and summing up the results to obtain the elements of the resulting matrix. This operation is crucial in various areas of mathematics, physics, and engineering.

Consider two matrices A and B:

$$A = \begin{bmatrix} 1 & 2 & 3 \\ 4 & 5 & 6 \end{bmatrix} \quad \text{and} \quad B = \begin{bmatrix} 7 & 8 \\ 9 & 10 \\ 11 & 12 \end{bmatrix}$$

To find the product AB, we multiply each element of the rows of A with the corresponding element of the columns of B, and sum up the results:

$$AB = \begin{bmatrix} 1 \cdot 7 + 2 \cdot 9 + 3 \cdot 11 & 1 \cdot 8 + 2 \cdot 10 + 3 \cdot 12 \\ 4 \cdot 7 + 5 \cdot 9 + 6 \cdot 11 & 4 \cdot 8 + 5 \cdot 10 + 6 \cdot 12 \end{bmatrix} = \begin{bmatrix} 7 + 18 + 33 & 8 + 20 + 36 \\ 28 + 45 + 66 & 32 + 50 + 72 \end{bmatrix}$$

$$= \begin{bmatrix} 58 & 64 \\ 139 & 154 \end{bmatrix}$$

Therefore, the product AB is a 2×2 matrix.

Matrix multiplication is not commutative, i.e., AB may not be equal to BA. It's important to note that the number of columns in the first matrix must be equal to the number of rows in the second matrix for the product to be defined.

Let's explore how to implement these computational rules using Python.

```
1    # Importing the needed library
2    import numpy as np
3
4    # Scalar x Vector
5    a = 10
6    v = np.array([1,2,3])
7    w = v*a
8
9    # Vector x Vector (element by element)
10   v1 = np.array([1,2,3])
11   v2 = np.array([4,5,6])
12   w  = v1*v2
13
14   # Vector + Vector (element by element)
15   w = v1+v2
16
17   # Scalar (inner) product
18   np.dot(v1, v2)
```

The last line of code calculates the scalar (inner) product of two vectors, v1 and v2, using NumPy dot() function. The scalar product, also known as the dot product, of two vectors is the sum of the products of their corresponding components. In mathematical terms, if v1 and v2 are both n-dimensional vectors: $scalar\_product = v1[0] * v2[0] + v1[1] * v2[1] + ... + v1[n-1] * v2[n-1]$, where $v1[i]$ and $v2[i]$ are the elements of vectors v1 and v2 respectively at index i. The dot() function is used to implement the rows x columns product. The element (i,j) of the resultant matrix, indeed, is the result of the scalar product between the i row-vector by the j column-vector. Below an example.

```
1   # Creating two matrices
2   m1 = np.array([[1, 2, 3], [4, 5, 6]])   # Matrix with 2
    ↪   rows and 3 columns
3   m2 = np.array([[7, 8], [9,10], [11, 12]])   # Matrix with
    ↪   3 rows and 2 columns
4
5   # Multiplying the matrices
6   result_matrix = np.dot(m1, m2)
```

2.1.3   Eigenvalues

To discuss eigenvalues, it is necessary to first define the concepts of vector space and linear dependence and independence.

2.1.4   Vector Space

A *vector space* is a set V along with two operations: vector addition and scalar multiplication. The set V and these operations must satisfy the following properties:

1. Closure under addition: If $\mathbf{u}, \mathbf{v} \in V$, then $\mathbf{u} + \mathbf{v} \in V$.
2. Closure under scalar multiplication: If c is a scalar and $\mathbf{v} \in V$, then $c\mathbf{v} \in V$.
3. Associativity of addition: $(\mathbf{u} + \mathbf{v}) + \mathbf{w} = \mathbf{u} + (\mathbf{v} + \mathbf{w})$ for all $\mathbf{u}, \mathbf{v}, \mathbf{w} \in V$.
4. Commutativity of addition: $\mathbf{u} + \mathbf{v} = \mathbf{v} + \mathbf{u}$ for all $\mathbf{u}, \mathbf{v} \in V$.
5. Existence of an additive identity: There exists an element $\mathbf{0} \in V$ such that $\mathbf{v} + \mathbf{0} = \mathbf{v}$ for all $\mathbf{v} \in V$.
6. Existence of additive inverses: For each $\mathbf{v} \in V$, there exists an element $-\mathbf{v} \in V$ such that $\mathbf{v} + (-\mathbf{v}) = \mathbf{0}$.
7. Distributivity of scalar multiplication with respect to vector addition: $c(\mathbf{u} + \mathbf{v}) = c\mathbf{u} + c\mathbf{v}$ for all scalars c and all $\mathbf{u}, \mathbf{v} \in V$.
8. Distributivity of scalar multiplication with respect to field addition: $(c + d)\mathbf{v} = c\mathbf{v} + d\mathbf{v}$ for all scalars c, d and all $\mathbf{v} \in V$.
9. Compatibility of scalar multiplication: $c(d\mathbf{v}) = (cd)\mathbf{v}$ for all scalars c, d and all $\mathbf{v} \in V$.
10. Identity element of scalar multiplication: $1\mathbf{v} = \mathbf{v}$ for all $\mathbf{v} \in V$, where 1 is the multiplicative identity in the field of scalars.

2.1.5   Linearly Dependent and Independent Vectors

Linearly dependent vectors are vectors in a vector space such that at least one of the vectors can be expressed as a linear combination of the others. Formally, a set of vectors $\{\mathbf{v}_1, \mathbf{v}_2, \ldots, \mathbf{v}_n\}$ is linearly dependent if there exist scalars $c_1, c_2, \ldots, c_n$, not all zero, such that:

$$c_1 \mathbf{v}_1 + c_2 \mathbf{v}_2 + \cdots + c_n \mathbf{v}_n = \mathbf{0}.$$

Conversely, vectors are *linearly independent* if none of them can be written as a linear combination of the others. A set of vectors $\{\mathbf{v}_1, \mathbf{v}_2, \ldots, \mathbf{v}_n\}$ is linearly independent if the only scalars $c_1, c_2, \ldots, c_n$ that satisfy the equation above are $c_1 = c_2 = \cdots = c_n = 0$.

2.1.6   Intuitive and Formal Definition of Eigenvalues

Intuitive Definition An eigenvalue is a special number associated with a matrix that gives information about the matrix's scaling effect on certain directions in the vector space. When a matrix acts on one of its eigenvectors, the output vector is simply a scaled version of the input vector. This scaling factor is the eigenvalue.

Formal Definition Let A be a square matrix and $\mathbf{v}$ be a non-zero vector. If there exists a scalar λ such that

$$A\mathbf{v} = \lambda \mathbf{v} \tag{2.1.1}$$

then λ is called an *eigenvalue* of A and $\mathbf{v}$ is the corresponding *eigenvector*.

2.1.7   The Importance of Eigenvalues in Ordinary Differential Equations

The concept of eigenvalues (or characteristic values) is fundamental in the realization of computational models using ordinary differential equations (ODEs) for several reasons:

- **System Stability**: The eigenvalues of a matrix associated with a system of linear ordinary differential equations (ODEs) determine the stability of the solutions. To assess stability, it is necessary to find the characteristic polynomial, which is obtained by computing $p(\lambda) = \det(A - \lambda I)$, where A is the system matrix and I is the identity matrix. Solving this polynomial provides the eigenvalues. If all eigenvalues have negative real parts, the system is asymptotically stable, meaning that the solutions approach an equilibrium point over time. Conversely, if any eigenvalue has a positive real part, the system is deemed unstable.

- **System Dynamics**: Eigenvalues describe the temporal dynamics of the system. Complex eigenvalues with non-zero imaginary parts indicate the presence of oscillations. The frequency of the oscillations is given by the imaginary part of the eigenvalue, while the rate at which these oscillations dampen or amplify is determined by the real part.

- **Model Reduction**: In many applications, it is possible to reduce the dimensionality of a complex system by focusing on the most significant modes associated with the most important eigenvalues. This technique, known as spectral decomposition or model reduction, allows simplifying the problem while retaining the essential characteristics of the system.

- **Numerical Solution**: Numerical methods for solving ODEs, such as the Runge-Kutta method or the Euler method, often require eigenvalue analysis to ensure numerical stability. The choice of the time step in these methods can depend on the distribution of the system eigenvalues.

- **Perturbation Analysis**: Eigenvalues help understand how a system responds to perturbations. The sensitivity of a system to parameter variations can be analyzed by examining how the eigenvalues change in response to small perturbations in the system matrix.

- **Asymptotic Properties**: Eigenvalues determine the asymptotic behavior of solutions to a system of ODEs. For example, in a linear system, the general solution can be expressed as a combination of exponentials of the form $\exp(\lambda_i t)$, where λ_i are the eigenvalues of the system.

2.2 Ordinary Differential Equations (ODEs)

2.2.1 Derivatives and Integrals

Imagine you're driving a car along a road, and you want to know how fast your speed is changing at any given moment. The derivative is like a tool that helps you figure that out. Think of your car speedometer. It tells you how fast you're going at any particular moment. Now, let's say you glance at your speedometer at different times. Each time you check, you get a different speed. The derivative is like looking at how much your speed changes when you check it really quickly. It tells you how quickly your speed is changing. Mathematically, if there is a function that describes how something changes over time or space, the derivative of that function tells you how much the function is changing at a specific point. It's like zooming in and seeing the slope of the function at that point.

Now, let's talk about the integral. If the derivative helps understand how things are changing at a specific moment, the integral helps understand the total accumulation of that change over a period of time or space. Imagine your car again. If you want to know how far you've traveled over a certain time period, you would add up all the little distances you covered at each moment. The integral is like this sum. It accumulates all those tiny changes to give you the total amount of change. So, if the derivative is about the rate of change, the integral is about the total change.

Derivatives and integrals are crucial for defining differential equations. Differential equations involve functions and their derivatives, showing how a function changes over time or space. These equations model real-world phenomena where change is involved. For instance, they can describe how a population grows, how heat diffuses through a material, or how a planet moves through space. Derivatives provide an understanding of the rate of change of these phenomena, and integrals provide an understanding of the accumulated effect of these changes. Together, derivatives and integrals allow the creation and solving of differential equations, providing deep insights into the dynamic systems around us.

Given a function $f(x) : \mathbb{R} \to \mathbb{R}$, the derivative $f'(x)$ measures how $f(x)$ changes with respect to x. The most common definition of a derivative is expressed using the *difference quotient*, which represents the rate of change of the function at a point:

$$Df(x) = f'(x) = \lim_{h \to 0} \frac{f(x+h) - f(x)}{h} \qquad (2.2.2)$$

Here:

- x is the point at which the derivative is evaluated.
- h represents a small increment in x, approaching 0.
- The expression $\dfrac{f(x+h) - f(x)}{h}$ approximates the slope of the tangent to the function at x.

The integral, which represents the accumulation of values of $f'(x)$ over an interval, can be described through the Riemann sum approximation. The Riemann integral is expressed as:

$$f(x) = \int_a^b f'(x)\, dx + c = \lim_{n \to \infty} \sum_{i=1}^{n} h f'(a + ih) + c \qquad (2.2.3)$$

In this equation:

- a and b are the limits of integration.
- n is the number of subintervals into which the interval $[a, b]$ is divided.
- $h = \frac{b-a}{n}$ is the width of each subinterval.
- The sum $\sum_{i=1}^{n} h f'(a + ih)$ approximates the area under the curve of $f'(x)$.
- c represents the constant of integration.

Notice that while conceptually integral and derivative can be thought as being the inverse function of each other, the inversion is not exactly symmetric as there is loss of information during the derivation process. This is expressed by the constant c in the integral formula. In fact, when computing the derivative of a function $g(x) = f(x) + c$, we obtain:

$$g'(x) = \lim_{h \to 0} \frac{g(x+h) - g(x)}{h} = \lim_{h \to 0} \frac{(f(x+h) + c) - (f(x) + c)}{h} = f'(x)$$

(2.2.4)

meaning that there are at least an infinity of functions $g(x)$ that share the same derivative function of $f(x)$, and the derivative function $f'(x)$ does not carry over any information about the constant c. When computing the integral of a function, therefore, it will be necessary to have at least some knowledge of the integral function in order to compute c and therefore, express the original integral function in terms of its derivative as $g(x) = \int g'(x)dx + c$.

2.2.2 Equations Versus Differential Equations

Equations and differential equations both serve as mathematical tools to relate variables to one another, yet they exhibit distinct characteristics. A conventional equation encompasses numerical values alongside at least one variable, denoted by a letter, representing the unknown within the equation. These equations typically employ algebraic operations like addition, subtraction, multiplication, and division to establish relationships between variables. Solutions to equations yield specific values or sets of values upon solving for the variable(s).

For instance, consider the equation $2x = 6$. The goal is to determine the numerical value that replaces the unknown x to render the equality true. In this case, the solution is not $x = 10$, as substituting x with 10 fails to satisfy the equation. The correct solution, where the equation holds true, is $x = 3$, exemplifying a scenario where only one value satisfies the equation. In equations involving multiple variables, such as $y = 2x$, an array of ordered pairs of numbers can satisfy the equation. For instance, if $x = 5$, $y = 10$ (5; 10) constitutes a solution, while if $x = 1$, $y = 2$ (1; 2) represents another solution, and so forth.

Differential equations, on the other hand, diverge from conventional equations in several significant ways. Notably, whereas traditional equations typically seek specific variable values or sets of values, differential equations necessitate *finding functions*. A function denotes a relationship between various quantities. For instance, one may consider a function, often graphically represented as a curve, correlating the height of a tree to the passage of time (Fig. 2.3).

This is an example of an increasing function but we can think of functions that have a different trend. Reality is full of functions, i.e. relationships between different elements. For example, the variation of the height of the lit candle over time has an opposite trend to that of the function representing the variation of the height of the tree over time. In fact, in the case of the candle the height decreases over time and the function is said to decrease. In the case of man, the curve that represents the trend of height over time is still different because it starts in an increasing way and then

Fig. 2.3 Growth of a tree height over time. The plot shows how the height increases each year

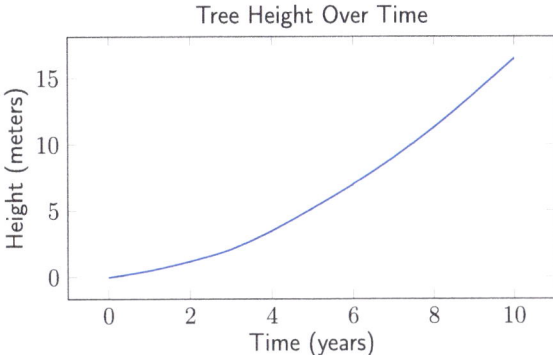

becomes decreasing. Thus, differential equations are equations whose solution is not a number but a function. When we solve a differential equation we get the curve representing that function.

A second notable distinction between traditional equations and differential equations lies in the inclusion of derivatives within the latter expressions. Unlike traditional equations, which typically involve direct relationships between variables, differential equations incorporate derivatives of the sought-after function. In doing so, they establish connections between variables through rates of change, as derivatives quantify how one variable evolves concerning another. Examples of differential equations are $\frac{dy}{dx} = x^2$ or $\frac{d^2y}{dx^2} + y = \sin(x)$.

ODEs are powerful tools used in various fields to model and solve problems involving rates of change. In physics, ODEs describe the motion of objects under forces, as exemplified by Newton's second law of motion. In biology, ODEs could model population dynamics, such as bacterial growth or disease spread, capturing how populations evolve. In economics, they help analyze and predict financial markets and economic growth, including the effects of investment strategies on economic development. Engineering relies on ODEs to design and analyze systems like electrical circuits, detailing how voltages and currents change. In neuroscience, ODEs model neural activity and brain dynamics, such as the propagation of electrical impulses through neural networks or the interactions between different brain regions during cognitive tasks. These examples highlight the versatility and significance of ODEs in understanding and forecasting the behavior of complex systems across diverse disciplines.

2.2.2.1 Practical Example

Consider a system of linear ODEs:

$$\frac{d\mathbf{x}}{dt} = A\mathbf{x}$$

where A is a matrix. The eigenvalues λ_i of A influence the solution of the system. If $\mathbf{v}_i$ is an eigenvector associated with λ_i, the general solution can be written as:

$$\mathbf{x}(t) = \sum_i c_i \mathbf{v}_i e^{\lambda_i t}$$

where c_i are constants determined by the initial conditions. The nature of the eigenvalues λ_i (positive, negative, imaginary) determines the behavior of the solution over time.

In summary, eigenvalues provide crucial information about the stability, dynamics, reducibility, and numerical solvability of models based on ordinary differential equations, making them an indispensable tool in computational modeling.

Now, let's see how we can calculate eigenvalues and eigenvectors using Python with the help of the NumPy library.

```python
import numpy as np
```

```python
# Define a 2x2 matrix
A = np.array([[4, 2],
              [1, 3]])

# Calculate eigenvalues and eigenvectors
eigenvalues, eigenvectors = np.linalg.eig(A)

print("Eigenvalues:")
print(eigenvalues)

print("\nEigenvectors:")
print(eigenvectors)
```

```
Eigenvalues:
[5. 2.]

Eigenvectors:
[[ 0.89442719 -0.70710678]
 [ 0.4472136   0.70710678]]
```

```
1   # Define a 3x3 matrix
2   B = np.array([[6, 2, 1],
3                 [2, 3, 1],
4                 [1, 1, 1]])
5
6   # Calculate eigenvalues and eigenvectors
7   eigenvalues, eigenvectors = np.linalg.eig(B)
8
9   print("Eigenvalues:")
10  print(eigenvalues)
11
12  print("\nEigenvectors:")
13  print(eigenvectors)
```

```
1   Eigenvalues:
2   [7.28799214 2.13307448 0.57893339]
3
4   Eigenvectors:
5   [[ 0.86643225  0.49742503 -0.0431682 ]
6    [ 0.45305757 -0.8195891  -0.35073145]
7    [ 0.20984279 -0.28432735  0.9354806 ]]
```

2.3 Mathematics on the Computer: Errors and Their Sources

When mathematics interacts with the real world, things often become more compli-cated. Mathematical models of physical systems aim to represent the system proper-ties based on their *measured quantities*. However, measurement errors always exist, making the true value of the property unknowable. These measurement errors are divided into two components: *random* and *systematic*. Random errors cause measur-able values to be inconsistent when repeated measurements of a constant quantity are taken. These errors are always present in a measurement and result from unpre-dictable fluctuations in the readings of a measurement apparatus or the experimenter's interpretation of the readings. Systematic errors, on the other hand, are predictable and typically constant or proportional to the true value (Fig. 2.4). These errors usually stem from imperfect calibration of measurement instruments or flawed observation methods. Identifying their cause can sometimes lead to applying corrections.

In any case, every number resulting from a measurement has some degree of error. A meticulous experimenter will associate an error estimation with each measurement, declaring a quantity as $x \pm y$, where x is the center value of a normal distribution and y specifies the uncertainty on that value. A less refined (but often useful) approach

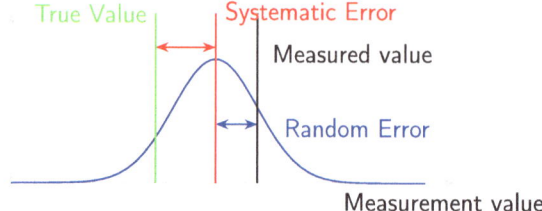

Fig. 2.4 Measurement errors. Systematic errors are constant in sign and magnitude, while random errors follow a probability distribution, centered around the true value dispaced by the systematic error

is to declare the precision of the measurement in terms of significant digits. For example, a measure of 123.42 ± 0.1 would have only three significant digits and should be approximated as 123 (or 124 if the decimal part were to be greater than 0.5). With this approach, any other number that interacts mathematically with this one should be truncated to the same number of significant digits.

Also in the realm of numbers errors can creep in. For a vast range of problems we do not have a mathematically exact solution: we do however have methods that allow us to compute an approximation of the solution, sometimes within an arbitrary error; this computational errors behave similarly to measurement errors.

Imagine to be solving a mathematical problem, which we will assume having an *exact* solution x, using a numerical method. The method will, in most cases, provide an *approximated* solution $\tilde{x}$.

The *absolute error* on the result, $\Delta x \equiv x - \tilde{x}$, while important, is usually not as useful as the *relative error* ϵ_x:

$$\epsilon_x \equiv \frac{\Delta x}{x} = \frac{x - \tilde{x}}{x} \tag{2.3.5}$$

The latter offers a representation that expresses the impact of the error itself on the value being represented: an error of 1 on 10^{10} is definitely less important than the same error on 10^1.

The sources of error are essentially of two kinds:

- due to the method: discretization errors (when infinite series have to be truncated to a finite number of terms, for example) and convergence errors (when an iterative method has to be stopped after a finite amount of iterations, for example)
- due to the arithmetic used: when using a finite arithmetic on a computer, rounding errors are likely to happen in every operation that is executed.

It is important to consider all sources of errors during development, and to assess their impact on the overall computation. It is usually possible, although usually complex, to perform a complete *error analysis* of a numerical method to assess how much errors are amplified by the method itself. The error analysis can be critical to understand if (and how much) a method's result can be trusted, and hence, it is a necessary instrument to choose the right method for the job. In general, the error analysis is synthesized in a coefficient κ which is called the *condition number* of the problem and expresses the amplification of the error from the input to the output of the method, such that:

$$|\epsilon_y| \approx \kappa |\epsilon_x| \tag{2.3.6}$$

Therefore, when $\kappa \approx 1$ the problem is said to be *well conditioned* and errors are not amplified; when instead $\kappa \gg 1$ the problem is said to be *ill conditioned* and will likely not produce a significant answer: one should try to find a different approach which is well conditioned instead.

2.3.1 Rounding Errors: Finite Arithmetic on the Computer

Numerical methods, regardless of the field in which they are defined (calculus, numerical analysis, linear algebra, differential equations and so on) are usually developed on continuous sets, like the set of real numbers $\mathbb{R}$ or complex numbers $\mathbb{C}$.

When using a computer, we have no choice but to represent any number x with a finite and constant amount of binary digits. In particular, numbers are normally represented following the IEEE754 standard [19] which divides the available bits (16, 32, 64, ...) in three parts:

- a sign s (normally 1 bit)
- an exponent $e = e_1, ..., e_m$ (8 bits for float32)
- a fractional part $f = f_1, ..., f_s$ (23 bits for float32).

such that, in a simplified form,[1] the number x is represented as:

$$x = s e_1 ... e_m f_1 ... f_s = (-1)^s \left(\sum_{i=1}^{s} f_i 2^{s-i} \right) 2^{r-v}, \quad r = \sum_{j=1}^{m} e_j 2^{m-j} \tag{2.3.7}$$

where usually $v = 2^{m-1}$. This is conceptually equivalent to the "E" scientific notation where a decimal number is expressed as $\pm x E y = \pm x \cdot 10^y$, with the fractional part and the exponent always using the same amount of digits.

Equation (2.3.7) shows that *machine representable numbers* have the following properties:

1. the set of numbers is *finite* (and has at most 2^n elements, where n is the amount of available bits (the standard defines 16, 32, 64, 128 and 256 bits representations)

---

[1] This formulation ignores many details and optimizations which are carefully defined in the standard, such as reserved combinations for special numbers (infinities, error conditions etc.) and normalized/denormalized fractional part, which allows to gain one significant bit when representing normalized numbers.

2. there is a smallest representable number (disregarding the sign) N_m which is in the order of 2^{-m+1} (in this simplified representation)
3. there is a biggest representable number (disregarding the sign) N_M which is in the order of 2^{m+s} (in this simplified representation)
4. the interval $[N_m, N_M]$ is not sampled uniformly, as there are approximately the same amount of numbers in the interval $[0, 1] \subset [N_m, N_M]$ as there are in $[1, N_M] \subset [N_m, N_M]$.
5. some numbers may have more than one representation (for example, 0 is also –0).

All the above implies that in general, a real number $x \in \mathbb{R}$ cannot be represented exactly, and we are forced to choose one of the nearest representations, introducing a *representation error* (or *rounding error*) every time. The IEEE754 standard defines in details the behaviour this approximations.

As an example of the difference between working in $\mathbb{R}$ or on computer, notice that properties such and associativity and distributivity of sum and multiplication stop being valid, because the order in which the approximations are done do have an effect on the final result of the operation. Also, since the interval is not uniformly sampled, the precision of a number depends on its magnitude. Let's consider a float16 number: the IEEE754 standard defines 1 bit for sign, 5 bits for exponent and 10 bits for the fraction. When the exponent has the maximum positive value (15), the fraction bits are representing numbers between 32768 and 65504 [20] and the precision is 32 (2^5), because we are 5 bits short from being able to represent the unit. Hence, the following happens:

```
from numpy import float16 as f
print('60000: %f' % f(60000))
print('60000+1: %f' % (f(60000)+f(1)))
print('60000+15: %f' % (f(60000)+f(15)))
print('60000+16: %f' % (f(60000)+f(16)))
print('60000+32: %f' % (f(60000)+f(32)))
print('60000+48: %f' % (f(60000)+f(48)))
print('60000+49: %f' % (f(60000)+f(49)))
```

```
60000: 60000.000000
60000+1: 60000.000000
60000+15: 60000.000000
60000+16: 60032.000000
60000+32: 60032.000000
60000+48: 60032.000000
60000+49: 60064.000000
```

This is of course correct and expected, and should serve as a reminder that the type of floats one uses should be chosen carefully depending on the size of numbers one needs to represent and the precision one expects.

The "take away" point of this section is to always keep in mind that numbers and mathematical operations, on a computer, do behave differently than one would expect when doing algebra on paper: not all numbers can be represented, the size of numbers matter, the order of operations matter and so on. Details and proofs are outside of the scope of this book; good starting points to delve deeper in the matter are of course the IEEE754 standard [19, 21–26].

It can be proven that multiplication and division are well conditioned operations. This is however not true for the sum: when summing two numbers which are really close to each other and of opposite sign ($x_1 \approx -x_2$), the condition number can grow arbitrarily large and lead to unexpected results [21, 24].

2.4 Finding Roots

One of the most important problems to be solved by numerical approximation using an iterative method is finding the root of an equation, namely, given a function $y = f(x)$, finding the value $\bar{x}$ such that $f(\bar{x}) = 0$.

The most basic iterative method to solve this problem is called *bisection method*. Consider a continuos function $f(x)$ on the interval $[a, b]$ such that $f(a)f(x) < 0$: since there is a change of sign, there must be a point $\bar{x} \in [a, b]$ such that $f(\bar{x}) = 0$. The bisection method proceeds as follows:

1. set $k = 0, [a_k, b_k] = [a_0, b_0] = [a, b]$, set tol as the desired precision for the result as an absolute error
2. set $c_{k+1} = a_k + \frac{1}{2}(b_k - a_k)$ (hence divide the interval in two equal parts $[a_k, c_{k+1}], [c_{k+1}, b_k]$)
3. if $|b_k - a_k| < tol$ or if $f(c_{k+1}) = 0$ stop and return c_{k+1}
4. if $f(a_k)f(c_{k+1}) < 0$ set $a_{k+1} = a_k, b_{k+1} = c_{k+1}$, else if $f(c_{k+1})f(b_k) < 0$ set $a_{k+1} = c_{k+1}, b_{k+1} = b_k$ (hence, select the interval that still contains the root)
5. set $k = k + 1$ and restart from step 3.

This method has the useful property of converging *globally*, as long as the initial conditions are respected it will arrive at a root within the set tolerance in a finite number of iterations. There are of course more complex and more efficient methods that can converge in a significantly smaller number of iterations, but may converge only locally (i.e., if the starting point is sufficiently near to the root) and may require additional prerequisites (i.e. f to be differentiable in the whole interval) [21–24]. This example is however sufficient to demonstrate the nature of a simple numerical method and some of the caveats that are common to all methods. In particular, notice that:

- $\bar{x}$ may not be representable in the finite arithmetic we are using, hence in general one cannot use $f(x) = 0$ as a stopping test. The solution could also be an irrational number, which can only be approximated with such a method by definition. There is therefore always an approximation error due to the method itself;

- the computation of c_{k+1} becomes ill-conditioned as the distance between a_k and b_k approaches zero, and the precision one can reach in finite arithmetic depends on the magnitude of a_k and b_k as we have seen in the previous section;
- the conditioning of the computation of $f(c_{k+1})$ depends on the complexity of the function f itself, and may get worse as the root is approached;
- the complexity of f may impose a limit on the number of iterations that can be performed given the computing power one has available, hence the precision one can reach may not be arbitrary but limited by time constraints;

Many root finding algorithms are already implemented in python's `scipy.optimize` and `numpy.roots` packages; it is unlikely that the average scientist would need to implement their own root finding method when there are such excellent implementations available. It is however important to have a solid idea of how each method performs, and choose wisely what algorithm to apply to best fit the problem at hand.

2.5 Linear Systems

Very often models involve the solution of linear systems of equations:

$$A\mathbf{x} = \mathbf{b} \tag{2.5.8}$$

where

$$\mathbf{x} = (x_1, ..., x_n)^T \in \mathbb{R}^n, \quad \mathbf{b} \in \mathbb{R}^n, \quad A = \begin{pmatrix} a_{11} & \cdots & a_{1n} \\ \vdots & & \vdots \\ a_{m1} & \cdots & a_{mn} \end{pmatrix} \in \mathbb{R}^{m \times n} \tag{2.5.9}$$

If A is nonsingular (hence formed of linearly independent vectors), then it is invertible ($\exists A^{-1}$ such that $A^{-1}A = I$) and therefore there exists a unique solution:

$$\mathbf{x} = A^{-1}\mathbf{b} \tag{2.5.10}$$

There are many libraries and tools available that implement matrix inversion and direct solution of linear systems in safe and efficient ways (for example, `linalg.solve` from the numpy library in python [27]), details on how to solve this problems are thoroughly presented in [21, 22, 24, 28]. It is important to keep in mind, however, that the condition number of a linear system depends on the matrix A and tends to grow as the matrix approaches being singular. Also notice that the cost of solving a linear system is, in general, cubic with its size ($O(n^3)$); if one needs to solve multiple problems that share the same large A matrix, computing and reusing its inverse A^{-1} (or other optimizations, like LU factorization [21, 22]) are computationally more efficient.

2.5.1 Nonlinear Systems

The bisection method seen in Sect. 2.4 (and in general, most of more refined root-finding methods) can be generalized to find an approximation of the solution of the multidimensional, non-linear problem:

$$F(\mathbf{y}) = \mathbf{0}, \quad F : J \subseteq \mathbb{R}^n \to \mathbb{R}^n \tag{2.5.11}$$

with a finite number of iterations, each involving the solution of linear systems [29, 30]. Many of this methods, as for the scalar case, are already implemented in the `scipy.optimize` library in python [31].

2.6 Approximation by Interpolation

It sometimes happen that a function is too computationally expensive to be evaluated many times, or that the function itself is not explicitly know at all (but it can be sampled at arbitrary points). In such cases, the function f can be approximated by means of simpler functions, usually polynomials, in a limited domain.

In particular, given a set of *nodes* $\{x_i | i = 0, ..., n\}$, we search for a polynomial $p_n(x)$ of degree $k \le n$ such that:

$$p_n(x_i) = f(x_i), \quad 0 \le i \le n, \quad x_i < x_j \, \forall i < j. \tag{2.6.12}$$

It can be proven that the interpolating polynomial exists and is unique; it can be expressed in several forms using different polynomial basis which have useful numerical properties. Working with polynomial approximations can also be useful to apply numerical methods that require the derivative of the function itself, in such cases where it is not know.

Note that the interpolation error, while being null by definition on the interpolation points, oscillates in between the interpolation points and grows exponentially (with the degree of the interpolating polynomial) outside of the interpolation interval extremes [21, 22, 24, 28]. The degree of the interpolating polynomial can be kept small by means of *piecewise* interpolation, where multiple polynomials are used to interpolate subsequent regions of the function. In particular, the interpolating function q is defined such that:

$$q(x_i) = f(x_i), \quad 0 \le i \le n, \quad x_i < x_j \, \forall i < j \tag{2.6.13}$$

where q is composed of r subsequent polynomials of degree k, connected in such a way to have continuos derivatives up to a chosen degree $m \le k$ over the whole interpolating function:

$$q \equiv \begin{cases} q_s \in \Pi_k & s = 0, ..., r - 1 \\ q_s(x_i) = f(x_i), & i = sk, ..., sk + k, \\ q_s^{(d)}(x_{sk+k}) = q_{s+1}^{(d)}(x_{(s+1)k}), & d = 0, ..., m \end{cases} \qquad (2.6.14)$$

Note that in this case the total number of interpolating points n must be an integer multiple of the number of interpolating functions r.

Make sure to check the `numpy.polynomial` and `scipy.interpolate` packages out [27, 31].

2.6.1   Least Squares Approximation

An important use case of interpolation is the definition of a function that best describes experimental data for which a function is not know (but can be hypothesized). Let us assume to have data in the form of discrete, possibly noisy and repeated, measurements:

$$(x_i, y_i), \quad i = 0, .., n. \qquad (2.6.15)$$

We will be searching for a function (usually a polynomial or a spline, but can also be more complex) that minimizes the error across all measurements; in particular, a common approach is to minimize the square errors.

Let $\mathbf{y} = (y_0, ..., y_n)^T$ be the vector of measurements on the points $\mathbf{x} = (x_0, ..., x_n)^T$, and $\mathbf{z} = f(\mathbf{x}) = (z_0, ..., z_n)^T$ the vector of the corresponding values of the approximating function f on the same set of points $\mathbf{x}$; the aim is therefore minimizing the quantity:

$$\|\mathbf{y} - \mathbf{z}\|_2^2 = \sum_{i=0}^{n} |y_i - z_i|^2. \qquad (2.6.16)$$

The computation of the function parameters that minimize (2.6.16) depends on the function itself (linear, polynomial, spline, non-linear etc.); for the linear and polynomial cases the optimal solution can be found solving linear systems, but finding the optimal parameters for non-linear function usually require more complex and costly iterative searches. Details can be found in [21–24, 32].

The following example compares three polynomial fits (linear, third and fifth degree) of some noisy data, using the least squares polynomial fit provided by `numpy`. Notice the difference between interpolating functions (which are, in effect, the models we choose to reproduce the data): using an appropriate model to approximate the data is fundamental.

```
1  import numpy as np
2  from scipy import interpolate
3  from matplotlib import pyplot as plt
4  np.random.seed(42)
5  x = np.linspace(0,3.5*np.pi,50)
6  y = np.sin(x) + np.random.random(len(x))-0.5
7
8  f1 = np.polynomial.Polynomial.fit(x,y,1)
9  f3 = np.polynomial.Polynomial.fit(x,y,3)
10 f5 = np.polynomial.Polynomial.fit(x,y,5)
11 fbase = np.linspace(0,3.5*np.pi,1000)
12 plt.figure()
13 plt.scatter(x,y, alpha=0.5, label='measured')
14 plt.plot(fbase,np.sin(fbase),'-.', alpha=0.5,
   ↪  label='real')
15 plt.plot(fbase,f1(fbase), label='1')
16 plt.plot(fbase,f3(fbase), label='3')
17 plt.plot(fbase,f5(fbase), label='5')
18 plt.grid()
19 _ = plt.legend()
```

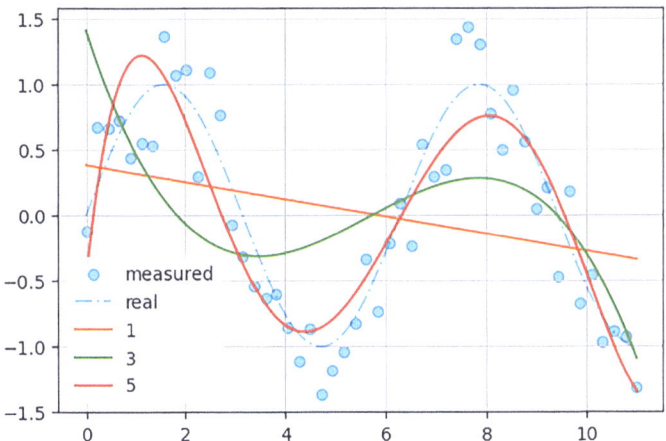

A fun example is the Anscombe's dataset, which provides examples that can be approximated linearly with exactly the same line and the same error:

```
1  x = [10, 8, 13, 9, 11, 14, 6, 4, 12, 7, 5]
2  y1 = [8.04, 6.95, 7.58, 8.81, 8.33, 9.96, 7.24, 4.26,
   ↪  10.84, 4.82, 5.68]
3  y2 = [9.14, 8.14, 8.74, 8.77, 9.26, 8.10, 6.13, 3.10,
   ↪  9.13, 7.26, 4.74]
4  y3 = [7.46, 6.77, 12.74, 7.11, 7.81, 8.84, 6.08, 5.39,
   ↪  8.15, 6.42, 5.73]
```

```
5
6   f1 = np.polynomial.Polynomial.fit(x,y1,1)
7   f2 = np.polynomial.Polynomial.fit(x,y2,1)
8   f3 = np.polynomial.Polynomial.fit(x,y3,1)
9   fbase = np.linspace(min(x), max(x), 1000)
10  fig, (ax1,ax2,ax3) = plt.subplots(1,3)
11  fig.set_size_inches(12,4)
12  for ax,f,y in [(ax1,f1,y1),(ax2,f2,y2),(ax3,f3,y3)]:
13      ax.scatter(x,y, alpha=0.5, label='Data')
14      ax.plot(fbase,f(fbase), color='r')
15      ax.grid()
```

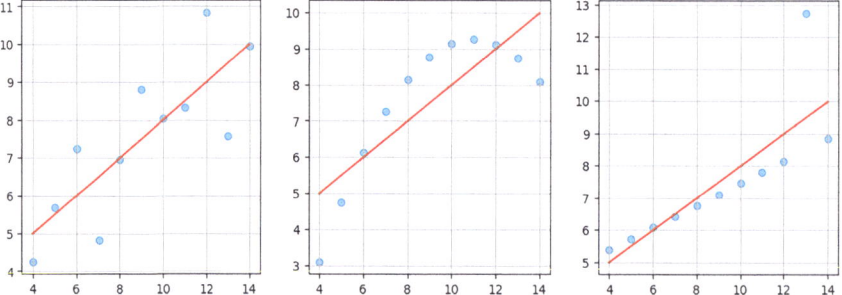

A second degree polynomial in this case already offers a much better representation of the data:

```
1   f1 = np.polynomial.Polynomial.fit(x,y1,2)
2   f2 = np.polynomial.Polynomial.fit(x,y2,2)
3   f3 = np.polynomial.Polynomial.fit(x,y3,2)
4   fbase = np.linspace(min(x), max(x), 1000)
5   fig, (ax1,ax2,ax3) = plt.subplots(1,3)
6   fig.set_size_inches(12,4)
7   for ax,f,y in [(ax1,f1,y1),(ax2,f2,y2),(ax3,f3,y3)]:
8       ax.scatter(x,y, alpha=0.5, label='Data')
9       ax.plot(fbase,f(fbase), color='r')
10      ax.grid()
```

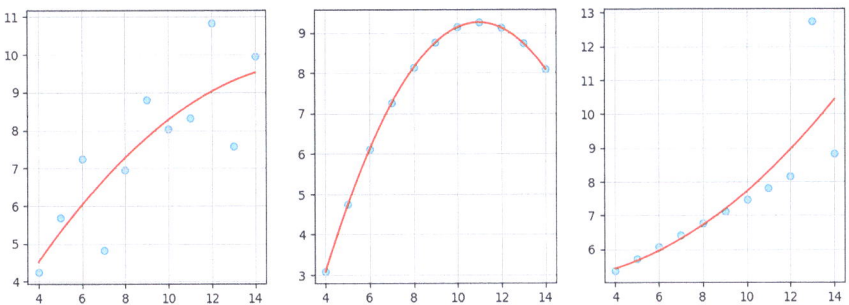

2.7 Difference Equations

The derivative of a continuous function $f(x)$ is typically defined according to Eq. 2.2.2, where x and h are continuous variables. As we have just seen continuous variables are not representable in a computer, therefore we have to work with approximations. In particular, our f can only be defined on a discrete set:

$$\Omega = \{x_0, x_1, x_2, ...\}, \quad x_n \in \mathbb{R}, \ x_n < x_{n+1} \tag{2.7.17}$$

where usually $x_{n+1} = x_n + h$ with h greater or equal than the smallest representable number, and can only assume the discrete values:

$$f_n \simeq f(x_n), \quad x_n \in \Omega. \tag{2.7.18}$$

which also need to be approximated to the nearest floating point number. In finite arithmetic a "limit going to zero" does not exist, but equations like (2.2.2) have a discrete counterpart, which can be expressed using the *difference operator* Δ for which:

$$\Delta f(x_n) = f(x_{n+1}) - f(x_n). \tag{2.7.19}$$

The theory of difference equations runs somewhat parallel to the theory of differential and integral calculus. For example, as in differential calculus we have that:

$$\int_{t=t_0}^{T} \frac{d}{dt} f(t) dt = f(T) - f(t_0) \tag{2.7.20}$$

it is straightforward to see that the equivalent holds:

$$\sum_{i=n_0}^{N-1} \Delta f(x_i) = (f(x_N) - f(x_{N-1})) + ... + (f(x_{n_1}) - f(x_{n_0})) = f(x_N) - f(x_{n_0})$$

$$\tag{2.7.21}$$

A differential equation of the first order is usually written in the form:

$$y'(x) = f(x, y(x)), \quad y' = \frac{dy}{dx} \tag{2.7.22}$$

which must be paired with an initial values for $y(x_0)$ to uniquely identify one of the solution functions $y(x)$. In general, real world problems are represented with higher order equations like:

$$y^{(n)} = f(x, y, y', y'', ..., y^{(n-1)}) \tag{2.7.23}$$

where[2] a solution function y is pinned by the initial values $y(x_0)$, $y'(x_0)$, ..., $y^{(n-1)}(x_0)$. In the discrete domain, our solution cannot be a continuous function $y(x)$, but will be a *sequence* y_{n_0}, ..., y_n, and the equivalent difference equation is therefore:

$$y_n = f(y_{n-1}, y_{n-2}, ..., y_{n-k}) \tag{2.7.24}$$

where k is the *order* of the difference equation; a solution sequence y_{n_0}, ..., y_n is also uniquely identified by the initial values y_{n_0}, ..., y_{n_0+k-1}.

When f is a linear combination, linear difference equations are generally represented in the form:

$$\sum_{i=0}^{k} p_i(n) y_{n+k-i} = g_n \tag{2.7.25}$$

where $\{p_i(n)\}$ and $\{g_n\}$ are known sequences. If p_i are constants, suche equations are called linear difference equations with constant coefficients.

If the equation is also homogeneous ($g_s \equiv 0$), it is possible to search for solutions in the form $y_n = z^{n-n_0}$, with $z \in \mathbb{C}$ and $n \geq n_0$, which results in the equation:

$$z^{n-n_0} \sum_{i=0}^{k} p_i z^{k-i} = 0 \tag{2.7.26}$$

therefore z must be one of the roots of the *characteristic polynomial* $p(z) = \sum_{i=0}^{k} p_i z^{k-i}$.

It can be proven that a general solution y_n can be constructed as a linear combination of k linearly independent solutions, and that such a base for the solution's space can be constructed using the roots z_1, ..., z_k, even in case they have higher multiplicity than 1; moreover, a general solution to the non homogeneous problem can be derived from such base in combination with a particular known solution.

Let's examine the stability of such solutions. Given a reference solution $\bar{y}_n$ for a linear difference equation like (2.7.25) and any other solution y_n, we can define the error sequence $e_n = \bar{y}_n - y_n$. Since $\bar{y}_n$, y_n are solutions of the same equation, e_n

---

[2] The (x) is usually omitted in the $y^{(n)}(x)$ expression for ease of notation, since it is clear from the context that y and all its derivatives are functions of x.

must be a solution of the associated homogeneous equation, and can be computed as linear combination of the roots of $p(z)$:

$$e_n = \sum_{i=1}^{k} \alpha_i z_i^{n-n_0} \qquad (2.7.27)$$

It is therefore evident that to have asymptotic stability, hence $lim_{n \to \infty} |e_n| = 0$, we must have $|z_i| < 1$. The case where $p(z)$ has roots with has a little more involved construction (and can be found in [33, 34]). The details are outside the scope of this introduction, although the final result is useful:

A solution y_n is:

- asymptotically stable if and only if $p(z)$ is such that $|z_i| < 1 \; \forall i$;
- stable if and only if $p(z)$ is such that $|z_i| \leq 1 \; \forall i$, and if $|z_i| = 1$ then z_i must be a simple root;
- unstable in any other case.

where:

- stable implies $|e_n| < \infty$ for $n \to \infty$;
- asymptotically stable implies $\lim_{n \to \infty} |e_n| = 0$
- unstable implies that $|e_n|$ is unbounded.

This summary is by no means complete and rigorous, but is hopefully sufficient to understand the arguments that follow; a more rigorous cover of those topics can be found in [21, 32–35].

2.8 Initial Value Problems

The problem:

$$y'(t) = f(t, y(t)), \quad y(t_0) = y_{t_0} \in \mathbb{R}^m \qquad (2.8.28)$$

is referred to as an Initial Value Problem since one can view the variable t as independent, and the equation as modeling a process that moves forward from some initial value t_0 with initial state y_{t_0}. In this continuous case, the function $y(t)$ describes a continuous trajectory which is the solution to the problem. Similarly, the recurrence:

$$y_{n+1} = f(n, y_n), \quad y_{n_0} \in \mathbb{R}^m \qquad (2.8.29)$$

identifies the corresponding discrete initial value problem, for which the discrete sequence $\{y_n\}$ is a solution.

Suppose that the initial value (2.8.28) is to be solved on the interval $[t_0, t_M]$, we divide this interval in uniform steps of size h such that:

$$t_n = t_0 + nh, \quad n = 0, \dots, N, \quad h = \frac{t_M - t_0}{N}, \quad N > 0 \tag{2.8.30}$$

we therefore transform the continuous problem (2.8.28) in the discrete one (2.8.29) by setting:

$$y_n = y(t_n), \quad f_n = f(t_n, y_n). \tag{2.8.31}$$

How to calculate (approximate) the solution's points sequence $\{y_n\}$? One of the simplest ways would be to use a formula of the form:

$$y_{n+1} = y_n + \Phi(t_n, y_n, h) \tag{2.8.32}$$

where Φ is a continuos function that ideally is exactly the integral of $f(t, y(t))$ in the interval $[t_n, t_{n+1}]$. But of course, unfortunately we don't know such a Φ, which would be the exact solution of (2.8.28): we'll have to approximate it. For example, Euler's method approximates f to be linear in each integration step (effectively approximating $y(t)$ with a sequence of line segments), and defines the recurrence to be:

$$y_{n+1} = y_n + h f_n \tag{2.8.33}$$

More in general, supposing to have more initial values to start with, one can think of combining those values linearly to obtain a better approximation. Given k initial values $y_0, \dots, y_{k-1}$, and the relation:

$$\sum_{i=0}^{k} \alpha_i y_{n+i} = h \sum_{i=0}^{k} \beta_i f_{n+i} \tag{2.8.34}$$

out of which we can extrapolate the unknowns:

$$\alpha_k y_{n+k} - h\beta_k f_{n+k} = -\sum_{i=0}^{k-1} \alpha_i y_{n+i} + h \sum_{i=0}^{k-1} \beta_i f_{n+i} \tag{2.8.35}$$

we have a general formula that defines a class of methods, each identified by the $2(k+1)$ coefficients α and β. Since the combinations of known values are linear, this class of methods is called *linear multistep formula* (LMF). If $\alpha_k \neq 0$, (2.8.35) can be divided by α_k and the equation would be the same. If $\alpha_k = 0$, there would be no value in the recurrence since it couldn't be used to compute y_{s+k}; it is therefore convenient to impose $\alpha_k = 1$. If $\beta_k = 0$, y_{n+k} can be computed directly and the method is called *explicit*. If instead $\beta_k \neq 0$, (2.8.35) defines the equation:

$$y_{n+k} - h\beta_k f_{n+k} - g_n = 0 \tag{2.8.36}$$

which is generally a non-linear equation, where g_n is the right hand side of (2.8.35) that is a known quantity, and its solution y_{n+k} can be approximated using a root finding method (as described in Sect. 2.4). Since y_n cannot be computed directly but require the solution of an equation at each step, this methods are called *implicit*. Euler's method is of course a LMF method, explicit, of order $k = 1$ and coefficients $\alpha_1 = 1, \alpha_0 = -1, \beta_1 = 0, \beta_0 = 1$.

How well does any method approximate the exact solution $y(t)$? Any method will introduce a *truncation error* at every step:

$$\sum_{i=0}^{k} \alpha_i y(t_{n+i}) - h \sum_{i=0}^{k} \beta_i f(t_{n+i}, y(t_{n+i})) = \tau_{n+k} \qquad (2.8.37)$$

Notice that in this formulation, the formula is using the *exact* value $y(t_n)$ and not the approximated one y_n: τ_n is indeed exactly the error that is introduced by the method *at each step*, and is a characteristic of the method itself.

A numerical method is said to be *consistent* with the differential equation it approximates if the truncation error τ_n is such that:

$$\forall \epsilon > 0 \ \exists h(\epsilon) > 0 \text{ such that } |\tau_n| < \epsilon \text{ for } 0 < h < h(\epsilon), \ \forall n = 0, ..., N \quad (2.8.38)$$

It is possible to define methods that have constrains on the truncation error, such as having $\tau_n = O(h^p)$ when $h \to 0^+$ (hence, the truncation error can be constrained to be consistent with an opportune power the step size h; in this case the method is said to have *order of accuracy p*).

Let's examine the case of LMF methods. Let $u_n = y(t_n)$ (and hence, $u'_n = f(t_n, y(t_n)) = y'(t_n)$) for ease of notation, expanding τ_{n+k} using Taylor's series we obtain:

$$\tau_{n+k} = \sum_{i=0}^{k} \alpha_i u_{n+i} - h \sum_{j=0}^{k} \beta_i u'_{n+i} \qquad (2.8.39)$$

$$= \sum_{i=0}^{k} \alpha_i \sum_{j \geq 0} \frac{u_n^{(j)}}{j!} (ih)^j - \sum_{i=0}^{k} \beta_i \sum_{j \geq 1} \frac{u_n^{(j)}}{(j-1)!} i^{j-1} h^j \qquad (2.8.40)$$

$$= u_n \sum_{i=0}^{k} \alpha_i + \sum_{j \geq 0} \frac{u_n^{(j)}}{j!} h^j \sum_{i=0}^{k} (i^j \alpha_i - j i^{j-1} \beta_i) \qquad (2.8.41)$$

from which it is evident that for any LMF method, if τ_n has to be $O(h^p)$, all the terms involving $h^l, l < p$ must vanish. Therefore, we must impose:

$$\sum_{i=0}^{k} \alpha_i = 0, \quad \sum_{i=0}^{k} (i^j \alpha_i - j i^{j-1} \beta_i) = 0, \quad j = 1, ..., p - 1 \qquad (2.8.42)$$

As $\tau_n \neq 0$, the errors at every step will sum up; it is therefore imperative to also consider the *global error*, defined as:

$$e_n = y(t_n) - y_n \tag{2.8.43}$$

which is the distance between the discrete solution points $\{y_n\}$ and their relative true values $y(t_n)$.

A method is said to be *convergent* if

$$\lim_{N \to \infty} \max_{n=0,\dots,N} |e_n| = 0 \tag{2.8.44}$$

or alternatively:

$$\lim_{n \to \infty} y_n = y(x) \text{ as } x_n \to x \in [x0, x_M] \text{ when } h \to 0 \text{ and } n \to \infty. \tag{2.8.45}$$

which means that it is possible to get arbitrarily close to the solution by reducing the step size h.

It is possible to prove that there exists methods such that if $\tau_n = O(h^p)$ and $e_i = O(h^r)$ for $i = 0, \dots, k - 1$ then $|e_n| \leq O(h^{\min(p,r)})$, $n = k, \dots, N$: in other words, it is possible to construct methods for which the global error stays of the same order of magnitude of the error on the initial conditions. This is really good news.

In the case of LMF methods, one can examine the global error by subtracting the definition of the method (2.8.34) from the truncation error as defined in (2.8.39) to obtain:

$$\tau_{n+k} = \sum_{i=0}^{k} \alpha_i (u_{n+i} - y_{n+i}) - h \sum_{j=0}^{k} \beta_i (u'_{n+i} - f_{n+i}) \tag{2.8.46}$$

$$= \sum_{i=0}^{k} \alpha_i e_{n+i} - h \sum_{j=0}^{k} \beta_i (y'(t_{n+i}) - f_{n+i}) \tag{2.8.47}$$

When considering a consistent method, for which $\tau_n = O(h^p)$, and since one assumes $h \to 0$ which implies that also $\tau_n \to 0$, it makes sense to ignore the right hand side of (2.8.46) and only consider the equation:

$$\sum_{i=0}^{k} \alpha_i e_{n+i} = 0 \tag{2.8.48}$$

to get a sense of the convergence properties of the method. This is actually a constant coefficients linear difference equation, and as was seen in Sect. 2.7, has an asymptotic null solution if $p(z) = \sum_{i=0}^{k} e_i z^{k-i}$ has all its complex roots inside the unit circle, or at least the error is bound if the stability conditions are met. In this case, the method is said to be *0-stable*, and it is a sufficient condition to ensure convergence.

To better understand, let's look at a simple example. Consider the problem:

$$y'(x) = -2y(x) + 3e^x, \quad y(0) = 3$$

for which we know the solution:

$$y(x) = 2e^{-2x} + e^x$$

and hence one can use it as an example to better understand the behaviour of our numerical solution. Euler's method to approximate $y(x)$ implies the following recurrence:

$$y_{n+1} = y_n + h(-2y_n + 3e^x)$$

```python
import numpy as np
from numpy import e
from matplotlib import pyplot as plt
y_true = lambda x:  2*e**(-2*x) + e**x
y_next = lambda x,y,h: y+h*(-2*y + 3*e**x)

def euler(x0, y0, h, N):
    X = [x0]
    Y = [y0]
    for i in range(0,N):
        X.append(X[-1]+h)
        Y.append(y_next(X[-1], Y[-1], h))
    return X,Y

xmin, xmax, y0 = 0, 1, 3
x = np.linspace(xmin,xmax,50)
Y = y_true(x)
X_hvs,Y_hvs = euler(xmin,y0,0.01,int(xmax/0.01))
X_hs,Y_hs = euler(xmin,y0,0.05,int(xmax/0.05))
X_hm,Y_hm = euler(xmin,y0,0.1,int(xmax/0.1))
X_hl,Y_hl = euler(xmin,y0,0.15,int(xmax/0.15))

plt.plot(x,Y, label='True solution')
plt.plot(X_hl,Y_hl, 'o', label='h 0.15')
plt.plot(X_hm,Y_hm, 'x', label='h 0.1')
plt.plot(X_hs,Y_hs, '.', label='h 0.05')
plt.plot(X_hvs,Y_hvs, '+', label='h 0.01')
plt.grid()
_ = plt.legend()
```

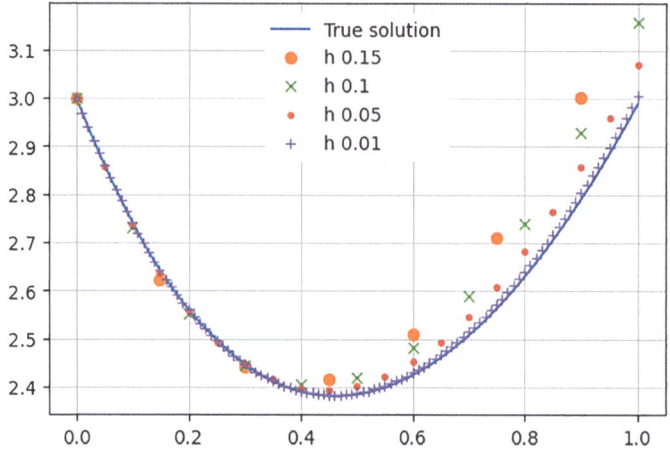

It is clearly visible how the step size h influences the truncation error τ_n and hence the global error e_n. According to (2.8.42) the method is consistent, hence one can expect to be able to get arbitrarily close to the solution by reducing the step size; however, depending on properties of the problem the step size h might need to become so small that the approxmation becomes computationally too expensive to be useful. Also, at each step, there is also a representation error being added, due to the very fact that we're using a machine with a limited number of bits to represent each number; performing too many steps may make the representation error more and more relevant, up to the point of losing any meaning (when the numbers are as small as the smallest number that can be represented in the machine, for example).

Obviously, one can do much better than Euler's method. Python's `scipy` library provides a function called `solve_ivp` which offers several advanced implementation of IVP itegrators. This integrators also implement explicit and implicit LMF methods, but apply advanced techniques to automatically identify the best (constant or variable) stepsize h, and to keep the method consistent and convergent whenever possible. The following example shows how `solve_ivp` with three different methods behaves on the same problem we just analyzed.

```
1   from scipy.integrate import solve_ivp
2   f = lambda x,y : -2*y + 3*e**x
3
4   res_bdf = solve_ivp(f, (0.0, 1.0), (3,), method='BDF')
5   res_lsoda = solve_ivp(f, (0.0, 1.0), (3,),
    ↪   method='LSODA')
6   res_RK23 = solve_ivp(f, (0.0, 1.0), (3,), method='RK23')
7
8   plt.plot(x,Y, label='True solution')
9   plt.plot(res_bdf.t, res_bdf.y[0], 'o', label='BDF')
10  plt.plot(res_lsoda.t, res_lsoda.y[0], 'x', label='LSODA')
11  plt.plot(res_RK23.t, res_RK23.y[0], '.', label='RK23')
12  plt.grid()
13  _ = plt.legend()
```

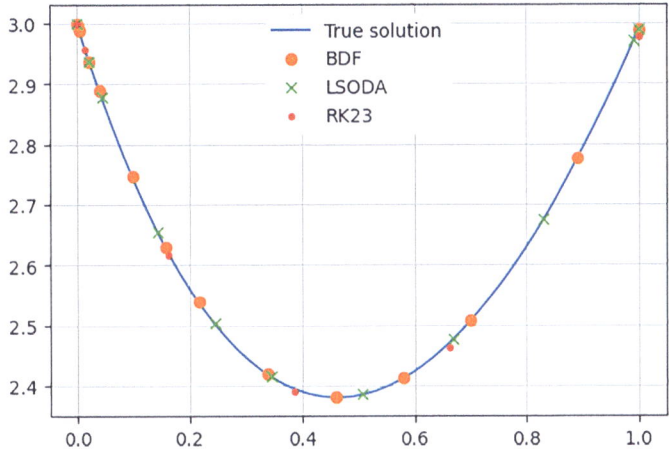

2.8.1   Stability with Fixed Step

Unfortunately, it is not always possible to count on reducing the stepsize to keep the error under control: on some problems, the required h could be so small that the method becomes in fact too computationally expensive to be used; or similarly, the required integration range could be very large (or infinite), making using an h as large as possible a practical necessity. It is therefore necessary to study the equation of the global error in the case where h is finite, and therefore the convergence and consistency properties are not enough to ensure the correctness of the result.

Let's apply Euler's method to a slightly different problem:

$$y'(x) = 10y(x) - 2y^2(x), \quad y(0) = 1, \quad x \geq 0$$

Also in this case, we know the exact solution:

$$y(x) = \frac{5}{1 + 4e^{-10x}}$$

```
1    import numpy as np
2    from numpy import e
3    from matplotlib import pyplot as plt
4    from scipy.integrate import solve_ivp
5
6    y_true = lambda x: 5./(1+4*e**(-10*x))
7    y_next = lambda x,y,h: y+h*(10*y-2*y**2)
8
9    xmin, xmax, y0 = 0, 3, 1
```

```
10   x = np.linspace(xmin,xmax,500)
11   Y = y_true(x)
12
13   def euler(x0, y0, h, N):
14       X = [x0]
15       Y = [y0]
16       for i in range(0,N):
17           X.append(X[-1]+h)
18           Y.append(y_next(X[-1], Y[-1], h))
19       return X,Y
20
21   X_hs,Y_hs = euler(xmin,y0,0.05,int(xmax/0.05))
22   X_hm,Y_hm = euler(xmin,y0,0.18,int(xmax/0.18))
23   X_hl,Y_hl = euler(xmin,y0,0.205,int(xmax/0.205))
24
25   plt.plot(x,Y, label='True solution')
26   plt.plot(X_hs,Y_hs, '.', label='h 0.05')
27   plt.plot(X_hm,Y_hm, 'x', label='h 0.18')
28   plt.plot(X_hl,Y_hl, '+', label='h 0.205')
29   plt.xlabel('x')
30   plt.ylabel('y', rotation='horizontal')
31   plt.grid()
32   _ = plt.legend()
```

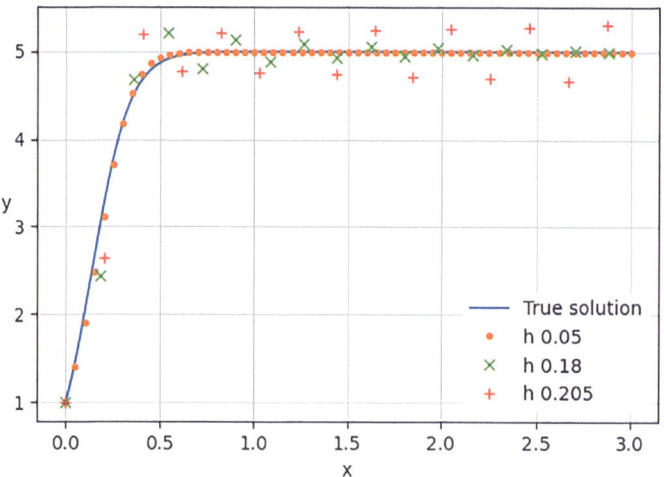

On this particular problem, Euler's method oscillates around the solution, with the oscillations becoming bigger with the stepsize. If one wouldn't have know the exact solution (which is usually the case), how could one determine that in fact, the oscillations aren't part of the solution? A further increase of the stepsize dramatically worsens the situation; the oscillations can become quite complex (as in chaotic):

```
1   xmin, xmax, y0 = 0, 30, 1
2   x = np.linspace(xmin,xmax,500)
3   Y = y_true(x)
4   X_hL,Y_hL = euler(xmin,y0,0.27,int(xmax/0.27))
5   plt.plot(x,Y, label='True solution')
6   plt.plot(X_hL,Y_hL, '+', label='h 0.27')
7   plt.xlabel('x')
8   plt.ylabel('y', rotation='horizontal')
9   plt.grid()
10   _ = plt.legend()
```

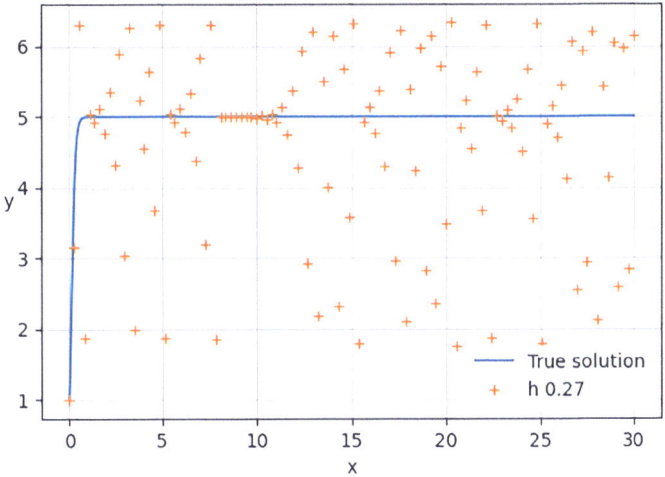

Or they can rapidly explode to nonsensically big numbers (notice the symmetric logarithmic scale in this graph):

```
1   xmin, xmax, y0 = 0, 3.5, 1
2   x = np.linspace(xmin,xmax,500)
3   Y = y_true(x)
4   X_hL,Y_hL = euler(xmin,y0,0.3002,int(xmax/0.3002))
5   plt.plot(x,Y, label='True solution')
6   plt.plot(X_hL,Y_hL, '+', label='h 0.4')
7   plt.yscale('symlog')
8   plt.xlabel('x')
9   plt.ylabel('y', rotation='horizontal')
10   plt.grid()
11   _ = plt.legend()
```

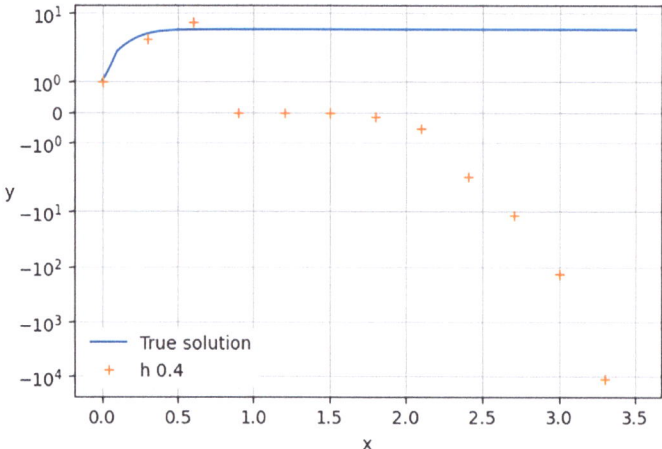

Especially when one doesn't knot much about the expected behaviour of the solution, which is the most common case, it is evidently really important to at least know if the method one will be applying can be expected to perform well.

A seemingly naive, but rigorously mathematically proven to be sound, way to get an idea of the behaviour of a method is to study its properties when applied to the test equation:

$$y'(x) = \lambda y(x), \quad y(0) = y_0, \quad \text{Re}(\lambda) < 0 \tag{2.8.49}$$

the solution of which is:

$$y(x) = y_0 e^{\lambda x} \longrightarrow 0, \quad x \to \infty \tag{2.8.50}$$

which means the origin is an asymptotically stable equilibrium point. Evidently, if also the discrete solution y_n is such that $y_n \to 0$ when $n \to \infty$, then it must be that the global error $e_n \to 0$ when $n \to \infty$. In the case of LMF methods, y_n is such that, as we have seen in (2.8.35):

$$\sum_{i=0}^{k} \alpha_i y_{n+i} = h \sum_{i=0}^{k} \beta_i f_{n+i} \tag{2.8.51}$$

which applied to the test Eq. (2.8.49) becomes:

$$\sum_{i=0}^{k} (\alpha_i - h\lambda\beta_i) y_{n+i} = 0 \tag{2.8.52}$$

for which, the corresponding error equation is:

$$\sum_{i=0}^{k} (\alpha_i - h\lambda\beta_i)e_{n+i} = \tau_{n+k}.$$ (2.8.53)

Considering $\tau_{n+k} = 0$, (2.8.53) is a constant coefficients difference equation, hence, as seen before, $e_n \to 0$ if and only if:

$$p(z) = \sum_{i=0}^{k} (\alpha_i - h\lambda\beta_i)z^i$$ (2.8.54)

is such that it has roots only strictly inside the unit circle. While the coefficients α_i and β_i are identified by the chosen method, h and λ are "free" parameters, where λ is set by the problem and h is really free for us to choose. We must therefore consider the parametric polynomial $p(z, h, \lambda)$, which is said to be the *stability polynomial* of the method. It is therefore useful to study the region:

$$\mathcal{D} = \{h \in \mathbb{R}^+, \lambda \in \mathbb{C}^- \text{ such that } p(z, h, \lambda) \text{ has only roots inside the unit circle}\}$$ (2.8.55)

as the method operating inside this region will show good asymptotic stability properties; a careful analysis of this region will finally allow one to choose the biggest step size h that keeps the method stable and hence reliable.

To better understand, let's examine Euler method's behaviour on the test equation. As we have seen before, the method's equation is:

$$y_{n+1} = y_n + hf_n$$ (2.8.56)

hence we have $\alpha_0 = -1, \alpha_1 = 1, \beta_0 = 1, \beta_1 = 0$ and the stability polynomial is therefore:

$$p(z, h, \lambda) = z - (1 + h\lambda)$$ (2.8.57)

for which the stability region can be obtained imposing:

$$|z| = |1 + h\lambda| < 1$$ (2.8.58)

and hence the method is stable only if h is sufficiently small, provided that $\lambda \in \mathbb{C}^-$.

In this particular case, the same condition can also be obtained by considering that:

$$y_{n+1} = y_n + hf_n = y_n + h\lambda y_n = (1 + h\lambda)y_n$$ (2.8.59)

and therefore:

$$y_n = (1 + h\lambda)^n y_0$$ (2.8.60)

which implies that the method is only conditionally stable, with the condition being $|1 + h\lambda| < 1$. Unfortunately the latter analysis is not so straightforward for all the methods.

Let's do a "back of the envelope" reasoning on the example we have seen in practice before: ignoring the nonlinear term (which only makes things worse), we had $\lambda \approx 10$, meaning that on this particular problem, any $h > 0$ would produce an unstable result when Euler's method is applied. In fact, even if it is not visible with the naked eye on the graph, also the solution with the smallest h never settles on 0 but keeps oscillating around the solution [21, 32–35].

2.9   Linear Systems of Equations

A single equation is rarely enough to describe a complex system; in most cases, a system of equations is necessary. In general, a system of differential equations of the first order can be represented as:

$$\mathbf{y}'(t) = A(t)\mathbf{y}(t) + \mathbf{b}(t), \quad t \geq t_0 \tag{2.9.61}$$

where the functions $\mathbf{y}, \mathbf{b} \in \mathbb{R}^m$ and $A(t) \in \mathcal{M}^{m \times m}$ for all $t \geq t_0$. Similarly to the one-dimensional case discussed earlier, $\mathbf{y}(t)$ describes an m-dimensional trajectory which is the solution to the problem, given an initial condition $\mathbf{y}_0$. The recurrence:

$$\mathbf{y}_{n+1} = A_n \mathbf{y}_n + \mathbf{b}_n, \quad n \geq n_0 \tag{2.9.62}$$

identifies the corresponding discrete problem, for which the sequence of vectors $\{\mathbf{y}_n\}$ is a solution. A general solution for (2.9.62) can be computed using the variations of constants method. Let's consider the associated homogeneous difference equation:

$$\mathbf{y}_{n+1} = A_n \mathbf{y}_n, \quad n \geq n_0 \tag{2.9.63}$$

and a sequence of matrices such that:

$$W_{n+1} = A_n W_n, \quad n \geq n_0 \tag{2.9.64}$$

If $\det(A_n) \neq 0$ for all $n \geq n_0$, then also:

$$\det(W_n) = \det\left(\prod_{i=n_0}^{n-1} A_i W_{n_0}\right) = \left(\prod_{i=n_0}^{n-1} \det(A_i)\right) \det(W_{n_0}) \neq 0 \tag{2.9.65}$$

hence W_n in invertible and we can find a solution given the initial condition $\mathbf{y}_{n_0}$ in the form:

$$\mathbf{y_n} = W_n \mathbf{c} = W_n W_{n_0}^{-1} \mathbf{y_{n_0}} = \Phi_{n,n_0} \mathbf{y_{n_0}}, \quad n \geq n_0 \tag{2.9.66}$$

Φ_{n,n_0} is called the *fundamental matrix* of the problem, and is the solution of:

$$\Phi_{n+1,n_0} = A_n \Phi_{n,n_0}, \quad n \geq n_0, \quad \Phi_{n_0,n_0} = I \tag{2.9.67}$$

The solution for the non-homogeneous problem (2.9.62) can be obtained by imposing $\mathbf{y}_n = W_n \mathbf{c}_n$ and the initial condition $\mathbf{y}_{n_0} = W_{n_0} \mathbf{c}_{n_0}$:

$$\mathbf{c}_{n_0} = W_{n_0}^{-1} \mathbf{y}_{n_0} \tag{2.9.68}$$

and, applying (2.9.62):

$$\mathbf{c}_{n+1} = \mathbf{c}_n + W_{n+1}^{-1} \mathbf{b}_n \tag{2.9.69}$$

we obtain:

$$\mathbf{c}_n = W_{n_0}^{-1} \mathbf{y}_{n_0} + \sum_{i=n_0}^{n-1} W_{i+1}^{-1} \mathbf{b}_i, \quad n \geq n_0 \tag{2.9.70}$$

from which, multiplying each side by W_n:

$$\mathbf{y_n} = W_n \mathbf{c_n} = \Phi_{n,n_0} \mathbf{y}_{n_0} + \sum_{i=n_0}^{n-1} \Phi_{n,i+1} \mathbf{b}_i, \quad n \geq n_0 \tag{2.9.71}$$

It can be prove that if $A_n \equiv A$, in other words if the system's coefficients are constant, then $\Phi_{n,s} = A^{n-s}$ and the solution becomes:

$$\mathbf{y}_n = A^{n-n_0} \mathbf{y}_{n_0} + \sum_{i=n_0}^{n-1} A^{n-i-1} \mathbf{b}_i \tag{2.9.72}$$

Finally, if also $\mathbf{b_i} \equiv \mathbf{b}$ and $1 \notin \sigma(A)$ then:

$$\bar{\mathbf{y}} = (I - A)^{-1} \mathbf{b} \tag{2.9.73}$$

is the *only* constant solution (or *equilibrium point*) of the equation.

It is possible to prove that:

$$\sum_{i=n_0}^{n-1} A^{n-i-1} = (I - A^{n-n_0})(I - A)^{-1} \tag{2.9.74}$$

and therefore:

$$\mathbf{y_n} = A^{n-n_0} \mathbf{y}_{n_0} + (I - A^{n-n_0})(I - A)^{-1} \mathbf{b} = A^{n-n_0}(\mathbf{y}_{n_0} - \bar{\mathbf{y}}) + \bar{\mathbf{y}} \tag{2.9.75}$$

This last equation allows to make some important considerations on the behaviour of the solution of the system:

$$\mathbf{y}_{n+1} = A\mathbf{y}_n + \mathbf{b} \tag{2.9.76}$$

where $\mathbf{y_0}$ is a given initial condition:

- if $1 \notin \sigma(A)$, there exist a unique equilibrium point $\bar{\mathbf{y}} = (I - A)^{-1}\mathbf{b}$
- if $|\sigma(A)| < 1$, meaning that all eigenvalues magnitudes are smaller than one, the equilibrium point is asymptotically stable
- if $\sigma(A)$ contains eigenvalues such that, if $|\lambda| = 1$ then it is *semisimple*, the the equilibrium point is stable
- if $\sigma(A)$ contains an eigenvalue with magnitude greater than one, the equilibrium point is unstable.

To better understand, let's look at a simple system of two equations:

$$\mathbf{y}' = A\mathbf{y}, \quad A \in \mathbb{R}^{2 \times 2} \tag{2.9.77}$$

where:

$$\mathbf{y} = \begin{pmatrix} y_1 \\ y_2 \end{pmatrix}, \quad A = \begin{pmatrix} a & b \\ c & d \end{pmatrix} \tag{2.9.78}$$

In this 2×2 case it's relatively easy to construct a matrix with given eigenvalues. In particular, we know that A's eigenvalues λ_1, λ_2 are the roots of its characteristic polynomial. Therefore, by imposing:

$$p(\lambda) = \lambda^2 - (a + d)\lambda + (ad - bc) = (\lambda - \lambda_1)(\lambda - \lambda_2) \tag{2.9.79}$$

we obtain the two constraint equations:

$$a + d = \lambda_1 + \lambda_2, \quad (ad - bc) = \lambda_1 \lambda_2 \tag{2.9.80}$$

which are two equations for four free parameters a, b, c, d. Let's impose two more arbitrary constraints:

$$b + c = \lambda_1 \lambda_2, \quad b - c = \lambda_1 + \lambda_2 \tag{2.9.81}$$

Now b, c can be computed solving the linear system:

$$\begin{pmatrix} 1 & 1 \\ 1 & -1 \end{pmatrix} \begin{pmatrix} b \\ c \end{pmatrix} = \begin{pmatrix} \lambda_1 \lambda_2 \\ \lambda_1 + \lambda_2 \end{pmatrix} \tag{2.9.82}$$

and in turn, by substitution, a, d:

$$a = \frac{\lambda_1 \lambda_2 + bc}{d}, \quad d = \frac{(\lambda_1 + \lambda_2) + \sqrt{(\lambda_1 + \lambda_2)^2 - 4(\lambda_1 \lambda_2 + bc)}}{2} \tag{2.9.83}$$

where we arbitrarily choose to use one of the two possible values for d, since it can only be found as the root of a second order equation.

```python
import numpy as np
from matplotlib import pyplot as plt

def compute_A(l1, l2):
    b,c = np.linalg.solve(np.matrix([[1,1],[1,-1]]),
        np.array([l1*l2,l1+l2]))
    d = ((l1+l2) + np.sqrt((l1+l2)**2-4*(l1*l2+b*c)))/2
    a = (l1*l2+b*c)/d
    return np.array([[a,b],[c,d]])

def solve(y0, A, b, n_steps):
    y = [y0]
    for i in range(n_steps):
        y.append( A.dot(y[-1]) + b)
    return np.array(y)
```

First of all, let's verify that our matrix building function works as expected:

```python
m = compute_A(-1,3)
print("Eigenvalues should be -1 and 3: %s " %
    np.linalg.eig(m)[0])
m = compute_A(0.023,-1.289)
print("Eigenvalues should be -1.289 and 0.023: %s " %
    np.linalg.eig(m)[0])
```

```
Eigenvalues should be -1 and 3: [-1.  3.]
Eigenvalues should be -1.289 and 0.023: [-1.289  0.023]
```

Now that we know how to build a matrix with given eigenvalues, we can observe the behaviour of a few systems. If both eigenvalues are such that $|\lambda| < 1$, we expect the solutions to converge to the equilibrium point that is defined by A and **b**. Here is the solution trajectory resulting from a diagonal matrix:

```
1  b, y0 = np.array([0,0]), np.array([1,1])
2  A = np.array([[0.9, 0], [0, 0.7]])
3  y = solve(y0, A, b, 100)
4  plt.scatter(y[:, 0], y[:, 1],
   ↪   c=plt.colormaps['viridis'].resampled(len(y)).colors)
5  plt.xlabel('$y_1$')
6  plt.ylabel('$y_2$', rotation='horizontal')
7  plt.grid()
```

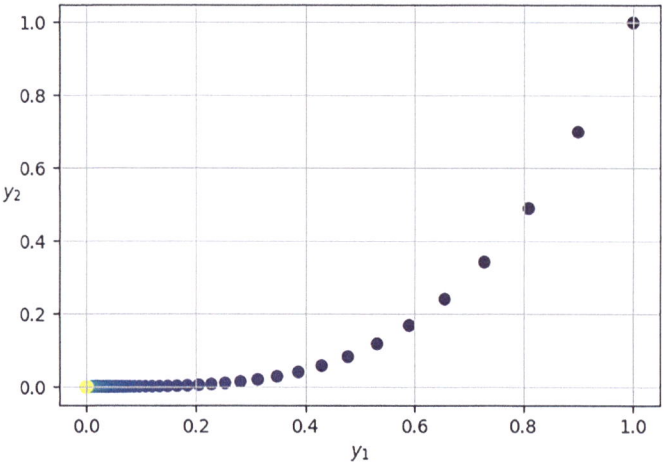

If the matrix is instead computed with the constraint we decided earlier, so that
the eigenvalues ar the same, the trajectory is different but the behaviour is still stable:

```
1  A = compute_A(0.9, 0.7)
2  print('A: ', A)
3  y = solve(y0, A, b, 100)
4  plt.scatter(y[:, 0], y[:, 1],
   ↪   c=plt.colormaps['viridis'].resampled(len(y)).colors)
5  plt.xlabel('$y_1$')
6  plt.ylabel('$y_2$', rotation='horizontal')
7  plt.grid()
```

```
1  A:  [[ 0.05785783  1.115      ]
2   [-0.485       1.54214217]]
```

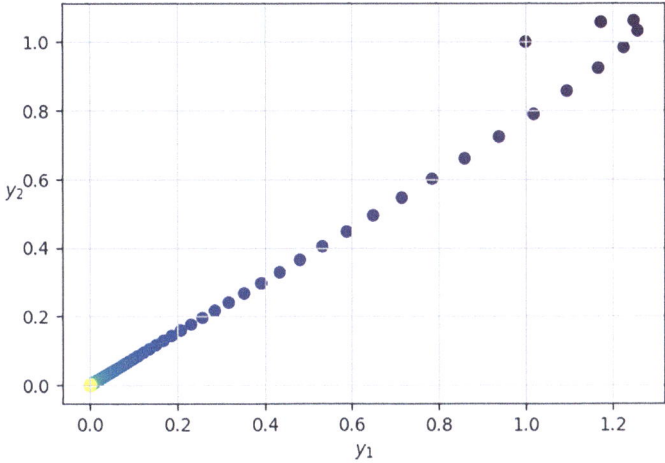

The choice of **b** does indeed change the equilibrium point:

```
1   b, y0 = np.array([2,5]), np.array([1,1])
2   A = np.array([[0.9, 0], [0, 0.7]])
3   y = solve(y0, A, b, 100)
4   plt.scatter(y[:, 0], y[:, 1],
    ↪    c=plt.colormaps['viridis'].resampled(len(y)).colors)
5   plt.xlabel('$y_1$')
6   plt.ylabel('$y_2$', rotation='horizontal')
7   plt.grid()
```

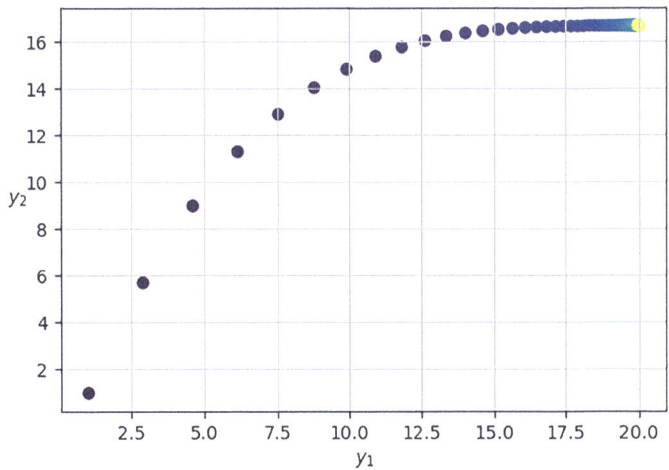

```
1   A = compute_A(0.9, 0.7)
2   y = solve(y0, A, b, 100)
3   plt.scatter(y[:, 0], y[:, 1],
    ↪   c=plt.colormaps['viridis'].resampled(len(y)).colors)
4   plt.xlabel('$y_1$')
5   plt.ylabel('$y_2$', rotation='horizontal')
6   plt.grid()
```

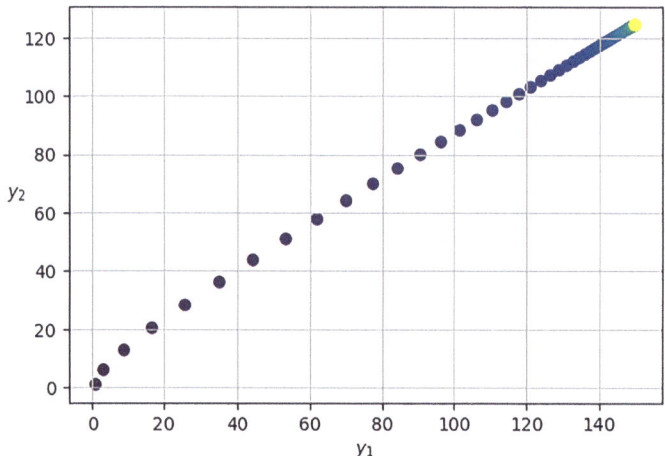

If instead the eigenvalues are greater that 1 in module, the solutions are unstable and diverge rapidly, as expected:

```
1   b, y0 = np.array([0,0]), np.array([1,1])
2   A = np.array([[1.1, 0], [0, 1.2]])
3   y = solve(y0, A, b, 100)
4   plt.scatter(y[:, 0], y[:, 1],
    ↪   c=plt.colormaps['viridis'].resampled(len(y)).colors)
5   plt.xlabel('$y_1$')
6   plt.ylabel('$y_2$', rotation='horizontal')
7   plt.grid()
```

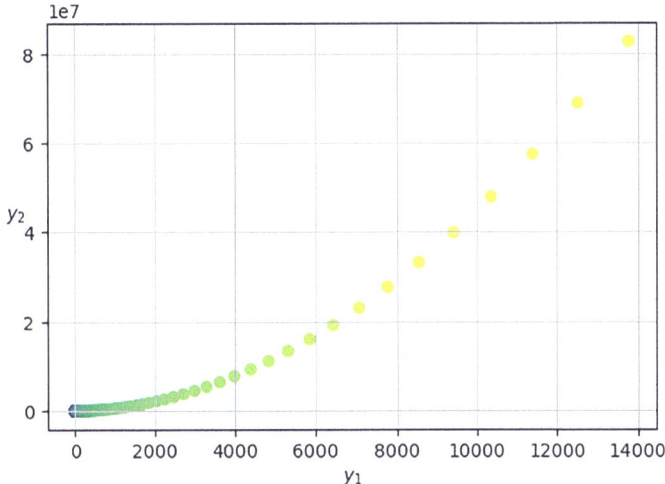

```
1  A = compute_A(1.1, 1.2)
2  y = solve(y0, A, b, 100)
3  plt.scatter(y[:, 0], y[:, 1],
   ↪  c=plt.colormaps['viridis'].resampled(len(y)).colors)
4  plt.xlabel('$y_1$')
5  plt.ylabel('$y_2$', rotation='horizontal')
6  plt.grid()
```

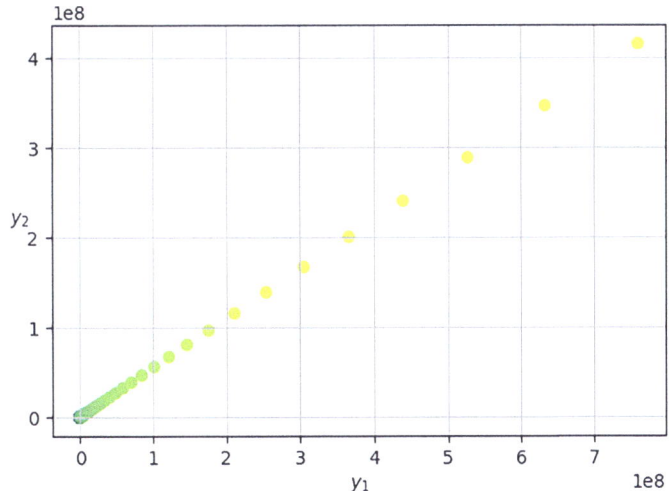

2.9.1   Higher Order Systems

The stability observations presented in Sect. 2.9 can be applied to higher order linear system of the kind:

$$\mathbf{y}_{n+k} = \sum_{i=1}^{k} A_i(n)\mathbf{y}_{n+k-i} + \mathbf{b}_n, \quad n \geq n_0 \qquad (2.9.84)$$

where $\mathbf{y}, \mathbf{b} \in \mathbb{R}^m$, $A_i(n) \in \mathbb{R}^{m \times m}$ as in fact, it is possible to express (2.9.84) as a first order system by constructing the block vectors:

$$\mathbf{u}_n = \begin{pmatrix} \mathbf{y}_n \\ \mathbf{y}_{n+1} \\ \vdots \\ \mathbf{y}_{n+k-1} \end{pmatrix}, \quad \mathbf{g}_n = \begin{pmatrix} 0 \\ 0 \\ \vdots \\ \mathbf{b}_n \end{pmatrix} \qquad (2.9.85)$$

and the block matrix:

$$H_n = \begin{pmatrix} O & I & & \\ & \ddots & \ddots & \\ & & O & I \\ A_k(n) & \cdots & A_2(n) & A_1(n) \end{pmatrix} \qquad (2.9.86)$$

and therefore expressing (2.9.84) in the form of (2.9.62):

$$\mathbf{u}_{n+1} = H_n \mathbf{u}_n + \mathbf{g}_n \qquad (2.9.87)$$

2.9.2   Stability

The same reasoning that has been described in Sect. 2.8.1 to analyze the stability of a scalar LMF method to solve an IVP can be applied, albeit with much more complex mathematics, to this vectorial case. While the detailed background theory and its application is outside the scope of this book, they can be found in [33, 36, 37]. In particular, a system of equations in the form:

$$\mathbf{y}'(t) = A\mathbf{y}(t) + \mathbf{b} = f(t, \mathbf{y}(t)), \quad t \geq t_0 \qquad (2.9.88)$$

can be solved with an LMF method in pretty much the same way previously discussed in Sect. 2.8, and the method's stability properties are identified by its characteristic polynomial.

An LMF method (as previously defined in (2.8.34)) applied to system (2.9.88) takes take the form:

$$\sum_{i=0}^{k} \alpha_i \mathbf{y}_{n+i} = h \sum_{i=0}^{k} \beta_i (A\mathbf{y}_{n+i} + \mathbf{b}) \tag{2.9.89}$$

To better understand the stabily properties of the method, we assume (2.8.34)) to have a unique stable equilibrium point $\bar{\mathbf{y}} = (-A^{-1}\mathbf{b})$; by translating the coordinates so that the equilibrium point is in the origin, the system becomes:

$$\mathbf{y}'(t) = A\mathbf{y}(t), \quad \sigma(A) \in \mathbb{C}^- \tag{2.9.90}$$

The system can now be transformed in the *Jordan normal form*:

$$\mathbf{y}'(t) = J\mathbf{y}(t) = \begin{pmatrix} J_1 & & \\ & \ddots & \\ & & J_k \end{pmatrix} \mathbf{y}(t) \tag{2.9.91}$$

where the transformed matrix J is composed of bidiagonal blocks (with only ones in the subdiagonal) or single-element diagonal blocks, depending on the multiplicity of each eigenvalue λ_i:

$$J_i = \begin{pmatrix} \lambda_i & & & \\ 1 & \lambda_i & & \\ & \ddots & \ddots & \\ & & 1 & \lambda_i \end{pmatrix} \in \mathcal{M}^{m_i \times m_i} \tag{2.9.92}$$

Since each Jordan block corresponds to a completely independent equation (or system of equations), the behaviour of each block can be studied independently. The case of unitary multiplicity is exactly the single equation case discussed earlier in Sect. 2.8.1. In the case of non-trivial Jordan blocks, the LMF method equation becomes:

$$\sum_{i=0}^{k} \alpha_i \mathbf{u}_{p,n+i} = h \sum_{i=0}^{k} \beta_i J_p \mathbf{u}_{p,n+i} \tag{2.9.93}$$

where $\mathbf{u}_{p,n}$ is the sub-vector of $\mathbf{y}_n$ corresponding to the elements that interact with the block J_p. Following the same reasoning that lead to (2.8.54) in the scalar case, for each Jordan block we obtain the vectorial polynomial:

$$p_i(\mathbf{z}) = \sum_{j=0}^{k} \left(\alpha_j I - h\beta_j J_i \right) \mathbf{z}^j \tag{2.9.94}$$

It can be proven that if all the polynomials:

$$p_i(z, h, \lambda_i) = \sum_{j=0}^{k} (\alpha_j - h\lambda_i \beta_j) z^j \qquad (2.9.95)$$

have all the roots strictly inside the unit circle (and is therefore a Schur polynomial), then also (2.9.94) is such that the method is asymptotically stable, hence $\mathbf{e_n} \to \mathbf{0}$.

2.10   NonLinear Systems of Equations

In general, the equations representing a system may not be linear. In such case, the system of differential equations (and its iterative counterpart) can be represented with a vectorial function:

$$\mathbf{y}'(t) = \mathbf{f}(t, \mathbf{y}(t)) \text{ with } t \geq t_0, \quad \mathbf{y_{n+1}} = \mathbf{f}(n, \mathbf{y_n}) \text{ with } n \geq n_0 \qquad (2.10.96)$$

The equilibrium point (or critical point) for the discrete problem $\mathbf{y_{n+1}} = \mathbf{f}(n, \mathbf{y_n})$ is $\bar{\mathbf{y}}$ such that:

$$\bar{\mathbf{y}} = \mathbf{f}(n, \bar{\mathbf{y}}) \qquad (2.10.97)$$

It can be assumed, without loss of generality, that $\bar{\mathbf{y}} = \mathbf{0}$; if that were not the case, the system can be translated by defining:

$$\mathbf{z}_n = \mathbf{y}_n - \bar{\mathbf{y}} \qquad \mathbf{z}_{n+1} = \mathbf{f}(n, \mathbf{z}_n + \bar{\mathbf{y}}) - \mathbf{f}(n, \bar{\mathbf{y}}) \equiv \mathbf{f}_t(n, \mathbf{z}_n) \qquad (2.10.98)$$

The null solution $\bar{\mathbf{y}}$ is defined as *stable* if:

$$\forall \epsilon > 0 \ \exists \delta = \delta(\epsilon, n_0) \text{ such that } \|\mathbf{y}_{n_0}\| < \delta \Rightarrow \|\mathbf{y}_n\| < \epsilon, \ \forall n \geq n_0 \qquad (2.10.99)$$

If δ does not depend on the state of the system, hence $\delta = \delta(\epsilon)$, the solution is *uniformly stable*. Furthermore, if $\lim_{n \to \infty} \mathbf{y}_n = \bar{\mathbf{y}} = \mathbf{0}$ the null solution is *asymptotically stable*. Uniform asymptotic stability ensure the existence of a distance from the null solution such that, once inside that range, any state of the system will converge to the null solution.

In most cases it is possible to see $f(n, \mathbf{y_n})$ as composed of a linear and a non-linear part, such that:

$$\mathbf{y_{n+1}} = f(n, \mathbf{y_n}) = A_n \mathbf{y_n} + \mathbf{g}(n, \mathbf{y_n}) \qquad (2.10.100)$$

where A_n is nonsingular and $\mathbf{g}(n, \mathbf{y_n})$ is, in general, non linear. By formally treating the nonlinear function $\mathbf{g}()$ as a given term, the solution $\mathbf{y}_n$ is, as seen in (2.9.71):

$$\mathbf{y_n} = \Phi_{n,n_0}\mathbf{y}_{n_0} + \sum_{i=n_0}^{n-1} \Phi_{n,i+1}\mathbf{g}(i, \mathbf{y}_i), \quad n \geq n_0 \tag{2.10.101}$$

This observation supports the intuition that the stability properties of the solution can be dominated by the linear part of the solution, under opportune hypotesis on the properties of $\mathbf{g}$.

The form (2.10.100) can be obtained, under the assumption that the function $\mathbf{f}$ is sufficiently regular, by applying Taylor's expansion up to the first term, so that:

$$\mathbf{f}(n, \mathbf{y}_n) = \mathbf{f}(n, \mathbf{0}) + \mathbf{J}_r(n, \mathbf{0})\mathbf{y}_n + \mathbf{g}(n, \mathbf{y}_n) = A_n\mathbf{y}_n + g(n, \mathbf{y}_n) \tag{2.10.102}$$

where $\mathbf{J}_r$ is the Jacobian matrix of $\mathbf{f}$ with respect to $\mathbf{y}$.

It can be proven that if $\mathbf{g}$ is such that:

$$\lim_{\|\mathbf{y}\|\to\mathbf{0}} \frac{\|\mathbf{g}(n, \mathbf{y})\|}{\|\mathbf{y}\|} = 0 \tag{2.10.103}$$

uniformly with respect to n, and also

$$\mathbf{y}_{n+1} = \mathbf{J}_r(n, \mathbf{0})\mathbf{y}_n \tag{2.10.104}$$

is uniformly asymptotically stable, then also the solution of the nonlinear system $\bar{\mathbf{y}} = \mathbf{f}(n, \bar{\mathbf{y}})$ is uniformly asymptotically stable.

It can be proven that this result also rigorously justifies the use of the test equation (2.8.49) $y'(x) = \lambda y(x)$ for the analysis the stability properties of numerical methods for solving ordinary differential equations near an asymptotic equilibrium point as seen in Sect. 2.8.1.

2.11 Numerical Optimization

Mathematical optimization is an open field of research, focused on finding algorithms and techniques to perform the selection of the best element, with regard to some criteria, from a set of available alternatives. In general, given:

- $\mathbf{x} \in \mathbb{D}^m$: a vector of variables (or unknowns); $\mathbb{D}$ represents any arbitrary set of values (usually it is the set of real or complex numbers);
- $f : \mathbb{D}^m \to \mathbb{R}$: the *objective* function, a scalar function defined on $\mathbb{D}^m$ which is to be minimized (or maximized);
- $c_1, ..., c_k : \mathbb{D}^m \to \{$False, True$\}$: a set of of scalar *constraint functions* c_i that define, through equations and inequalities, a set of criteria that $\mathbf{x}$ must satisfy;

the optimization problem is represented as:

$$\min_{\mathbf{x} \in \mathbb{D}^m} f(\mathbf{x}) \quad \text{subject to} \quad c_i(\mathbf{x}) = \text{True}, \ i = 1, .., k \qquad (2.11.105)$$

Optimization can be approached in many ways. How to choose the most efficient approach in general depends on the properties of the problem: whether the domain is continuous or discrete, if there are constraints, if one is searching for a global or a local optimum, if the approach has to be deterministic or can be stochastic, what are the convexity properties of the objective function, and so on. Minimization and maximization problems are evidently specular and can be transformed into one another by simply flipping the sign of f; we therefore only refer to minimization problems for the sake of simplicity.

In general, optimization algorithms are iterative: they begin with an initial guess of the solution $\mathbf{x}$ and they generate a sequence of estimates until some termination condition are met. A good optimization algorithm should exhibit at least three properties:

- Robustness: it should perform well on a wide variety of problems in the target class, with any reasonable value as a starting point;
- Efficiency: it should require limited storage and time;
- Accuracy: it should be able to identify a solution within the required precision without being sensitive to the accumulating numerical errors that inevitably happen in finite arithmetics.

The above properties may conflict: a strongly convergent algorithm may require too much storage space, or an accurate method may require too many iterations. In general, any approach identifies a tradeoff between this properties, and it is up to the user to have a firm grasp of the streghts and weaknesses of each algorithm to identify the one that may be best suited to solve a particular problem.

Great resources to start delving into the topic of numerical optimization are [38, 39].

2.11.1 Local and Global Optimization

A solution to the global optimization problem is the global minimizer $\bar{\mathbf{x}}$:

$$\bar{\mathbf{x}} \text{ such that } f(\bar{\mathbf{x}}) \le f(\mathbf{x}) \ \forall \bar{\mathbf{x}} \in \mathbb{D}^m \qquad (2.11.106)$$

However, such minimizer is seldom within reach: our knowledge about the function f is usually limited, local, and expensive to obtain. While in some cases (smooth mathematical surfaces) it may be possible to analytically prove the existence of a global minimizer, in general we cannot be sure that f does not take a sharp dip in some region that was not sampled by the optimization algorithm. Most appraches are therefore only able to identify local optimums $\mathbf{x}^*$ such that:

$$f(\mathbf{x}^*) \leq f(\mathbf{x}) \ \forall \ \mathbf{x} \in \mathcal{N} \tag{2.11.107}$$

where $\mathcal{N}$ is a limited neighborhood of $\mathbf{x}^*$ in $\mathbb{D}^m$. When f is sufficiently regular, analytical properties of the gradients of f can be used to discern if an identified point is indeed a local minima or is instead sitting in a region where the function is not convex.

Two Basic Strategies Serve as the basis for the approach implemented by the vast majority of the algorithms: *line search* and *trust region*. The line search strategy chooses a direction $\mathbf{p}_k$ and a step length α_k to move from the current position $\mathbf{x}_k$ so that:

$$\mathbf{x}_{k+1} = \mathbf{x}_k + \alpha_k \mathbf{p}_k, \quad \alpha_k \text{ such that } \min_{\alpha_k > 0} f(\mathbf{x}_k + \alpha_k \mathbf{p}_k) \tag{2.11.108}$$

In general (2.11.108) is too expensive to solve exactly, and usually that precision is not worth the cost; a limited number of samples and some interpolation strategy are generally enough to find a useful value for the current iteration step.

Trust region methods instead use a limited number of samples inside a carefully chosen neighborhood $\mathcal{N}_k$ of the current iteration step to build a *model function* $\mathbf{m}_k$ which is easily mathematically minimizable, and hence the method chooses the optimus of $\mathbf{m}_k$ as the candidate for the next iteration step:

$$\mathbf{x}_{k+1} = \mathbf{x} \text{ such that } \min_{\mathbf{x} \in \mathcal{N}_k} m_k(\mathbf{x}) \tag{2.11.109}$$

if $f(\mathbf{x}_{k+1}) \not\leq f(\mathbf{x}_k) - \delta$, meaning that the new candidate did not in fact find a better position within a required margin, the trust region $\mathcal{N}_k$ is modified (usually, reduced in size) and another step is performed.

2.11.1.1 Line Search

An example of line search method can be dirived from the Newton Method, which can be adapted from being a root-finding method to a corresponding optimization method. This method was originally formulated as a root finding method (see Sect. 2.4). Given a function $f(x)$, its tangent line in x_n that intercepts the x-axis has slope:

$$f'(x_n) = \frac{f(x_n) - 0}{x_n - x_{n+1}} \tag{2.11.110}$$

from which we can compute x_{n+1}:

$$x_{n+1} = x_n - \frac{f(x_n)}{f'(x_n)} \tag{2.11.111}$$

This latter equation is exactly the definition of the iterate for the root-finding Newton's method.

In this case, however, we are just interested in moving towards a lower point in the function (and we are not interested in wether it crosses the x-axis). Intuitively, $-\frac{f(x_n)}{f'(x_n)}$ is a good candidate direction to use in the iteration of the line search (2.11.108), which also implicitly contains information about how fast the object function is changing and therefore how big a step we should choose. Therefore the line search equation could become:

$$x_{k+1} = x_k - \alpha_k \frac{f(x_n)}{f'(x_n)} \qquad (2.11.112)$$

This idea can be extended to multidimensional domain, where the line search iteration becomes:

$$\mathbf{x}_{k+1} = \mathbf{x}_k - \alpha_k (\nabla f(\mathbf{x_k})^T)^{-1} f(\mathbf{x_k}) \qquad (2.11.113)$$

In general, of course, our knowledge of the properties of f may be quite limited; the Jacobian $\nabla f(\mathbf{x})$ (and higher order derivatives needed for more advanced methods) is usually unknown, but can be approximated with a finite difference, for example, while incurring in additional computational costs and approximation errors which may affect the needed number of iterations in significant ways. This extremely simplified description is of course ignoring many important aspects that influence the convergence, speed and stability properties (which can be found in [38]), but is hopefully sufficient to get a basic idea of how this kind of algorithms work.

2.11.1.2   Trust Region

A basic approach to the trust-region method is to construct the model function m_k as a truncated Taylor's approximation of f in the neighborhood of the current point x_k, for example, in the monodimentional case, m_k could be defined as:

$$m_k(x_k + h) = f(x_k) + f'(x_k)h + \frac{1}{2} f''(x_k)h^2 \qquad (2.11.114)$$

which is, therefore, a quadratic equation with respect to the *displacement* h. Let $H_k(x) : \mathbb{R} \to \mathbb{R} \times \mathbb{R}$ be the function that defines the trust region around a given point x inside which we are going to minimize m_k. In the simplest case, H would simply define a symmetric interval:

$$H_k(x) = (x - \alpha_k, x + \alpha_k) \qquad (2.11.115)$$

which then is used to solve the minimization problem that identifies the next step::

$$x_{k+1} = x \in H_k(x_k) \text{ such that } m_k(x) \leq m_k(z) \forall z \in H_k(x_k) \qquad (2.11.116)$$

which in this case, can be determined analytically. This approach translates directly to the multidimensional case, where:

$$m_k(\mathbf{x}_k + \mathbf{h}) = f(\mathbf{x}_k) + \nabla f(\mathbf{x_k})^T \mathbf{h} + \frac{1}{2}\mathbf{h}^T \nabla^2 f(\mathbf{x}_k)\mathbf{h} \qquad (2.11.117)$$

and for example, $H_k(\mathbf{x}_k)$ is defined as a sphere of radius α_k.

The trust radius α_k is updated at every step, usually in function of the agreement between the object function f and its model m_k: if the obtained reduction in $f(\mathbf{x}_{k+1}) - f(\mathbf{x}_k)$ matches the expected reduction in $m_k(\mathbf{x}_{k+1}) - m_k(\mathbf{x}_k)$, it makes sense to increase the trust radius to speed up the convergence. If instead the model is not matching well the object function f, the step should be reduced as we lose trust in the model function within this region.

2.11.2 Derivative-Free Optimization

The principles described in the previous sections, and all the actual methods that can be implemented following those principles, all require knowledge of the derivative (sometimes, also the second derivative and more) of the object function f. This can sometimes be a strong limitation: while gradients and Hessians can be approximated, the approximation process can be too computationally expensive, or completely impossible when the objective funtion is subject to *noise*, which in this context can be seen as just the result of inaccuracies in the function evaluation, but can also have other sources (for example, f could be the result of a stochastic simulation). One possible approach is to incrementally approximate f with a series of functions m_k which interpolate f on some of the previously sampled points $f(\mathbf{x_0})$, ..., $f(\mathbf{x_{k-1}})$, hence using the known trust region or line search algorithms on the interpolating functions to determine some approximation of x_k.

A commonly used approach is the *Nelder-Mead simplex-reflection* method. This method only use samples of f to determine a simplex[3] trust region to define the volume inside which to search for a better point that has a lower value of f, and that can therefore replace one of the vertices and shrink the trust region until some termination conditions are met. Practical performance of this algorithms is often reasonalble, and even though it may be subject to occasional stagnation, it is often worth trying it when faster methods cannot be applied.

---

[3] In geometry, a simplex is the multidimensional generalization of the concept of a triangle: a point, a line segment, a trangle, a tetrahedron, and so on.

2.11.3   Genetic Algorithms

When the objective function is partiulary complex, non differentiable, noisy, multi-modal, or the search space is really vast in terms of domain and number of dimensions, the approaches presented before may not be enough: they may be sensitive to the starting point (hence finding a different local minimum each time), or they may jump between different local minima wells if tolerances and step size are too big, they may not converge at all, and so on. Genetic algoritms apply the idea of gene recombination and survival of the fittest to the optimum search problem. Figure 2.5 shows the basic idea if such an algorithm: points $\mathbf{x}$ in the domain space are intepreted as representative individuals of a population of values of f, where the m dimensions of the domain are loosely interpreted as the genes that ultimately characterize the value of f; given a starting point $\mathbf{x}_0$, a population of candidates is generated by randomly mutating one of more components the vector, f is then evaluated on the candidate points, the best ones are kept, and a new generation of recombined and mutated individuals is put to test. There are many strategies that can be used to apply the random mutations, to recombine the population vectors and to choose which individuals to keep for the next iteration. Each strategy impact the convergence speed and the probability of finding a a good enough result once the region containing the global minimum has been found. Choosing the right strategy usually require careful testing and ranking on the problem that is being solved, so that the best balance of convergence speed and precision can be found. Genetic algorithms are relatively easy to understand, and technically easy to parallelize also in non-synchronous ways, at little cost with respect to convergence speed. The latter can be a great advantage over iterative methods that are difficult to parallelize, especially when f is really expensive to compute but can be computed in parallel by multiple machines. This algorithms can however be significantly more computationally expensive than advanced iterative methods, especially in cases where derivatives can be computed or approximated with sufficient precision.

Fig. 2.5 A flowchart representation of a generic genetic algorithm

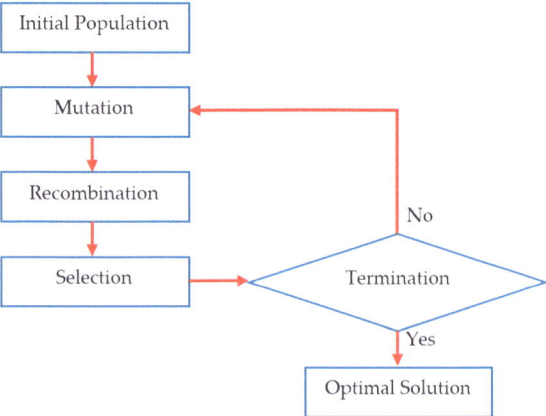

2.11.3.1 Differential Evolution

In [1] *differential evolution* (DE) is explained: it is a modern and reliable algorithm that can be applied and has been demonstrated to yield good results on a vast range of problems. An implementation is readily availble in python's `scipy` package [40].

DE is a genetically-inspired, population-based optimizer. The objective function is sampled at multiple, randomly chosen initial points (within the domain constraints) which form the initial population of N_p vectors. Similarly to other population-based methods (like Nelder-Mead for example), the population is replaced by new vectors which are perturbations and combinations of the existing ones in successive stages.

The most basic implementation of DE is described by the following pseudo-code, in which $\mathbf{x_i}$ is one of the N_p vectors of the population of one generation; this population will be replaced by the vectors being defined as $\mathbf{y_i}$ during the next generation. f is the fitness we want to maximize, and $F \in (0, 1)$ is called the *scale factor*:

```
1   while (convergence criterion not met) :
2       for i in range(N_p) :
3           r1 = random_int(0, N_p)
4           r2 = random_int(0, N_p)
5           r3 = random_int(0, N_p)
6           u_i = x_r3 + F * (x_r1 - x_r2)
7           if f(u_i) >= f(x_i) :
8               y_i = u_i
9           else:
10              y_i = x_i
```

This strategy of generating new vectors, called *mutation*, is usually paired with another one, called *crossover*, which acts on the individual components of the vectors. In particular, an intermediate candidate vector $\mathbf{v_i}$ is generated by mutation from three randomly selected members of the population $\mathbf{x_{r1}}, \mathbf{x_{r2}}, \mathbf{x_{r3}}$ as shown before in the pseudocode:

$$\mathbf{v_i} = \mathbf{x_{r3}} + F(\mathbf{x_{r1}} - \mathbf{x_{r2}}) \tag{2.11.118}$$

but additionally, the candidate vector for replacing $\mathbf{x_i}$, $\mathbf{u_i}$, has its components randomly selected from either $\mathbf{x_i}$ or $\mathbf{v_i}$:

$$(\mathbf{u_i})_j = \begin{cases} (\mathbf{v_i})_j, & \text{if } rand(0, 1) \leq C_r \\ (\mathbf{x_i})_j, & \text{otherwise} \end{cases} \tag{2.11.119}$$

where $C_r \in [0, 1]$ defines the *crossover probability*. $\mathbf{u_i}$ is then selected instead of $\mathbf{x_i}$ if $f(\mathbf{u_i}) >= f(\mathbf{x_i})$ as before. Evidently many details are being omitted for the sake of simplicity: at the very least, it is important to ensure that the three vectors involved in the mutation are in fact distinct from each other and that the new candidate vector has all its components within the allowed bounds.

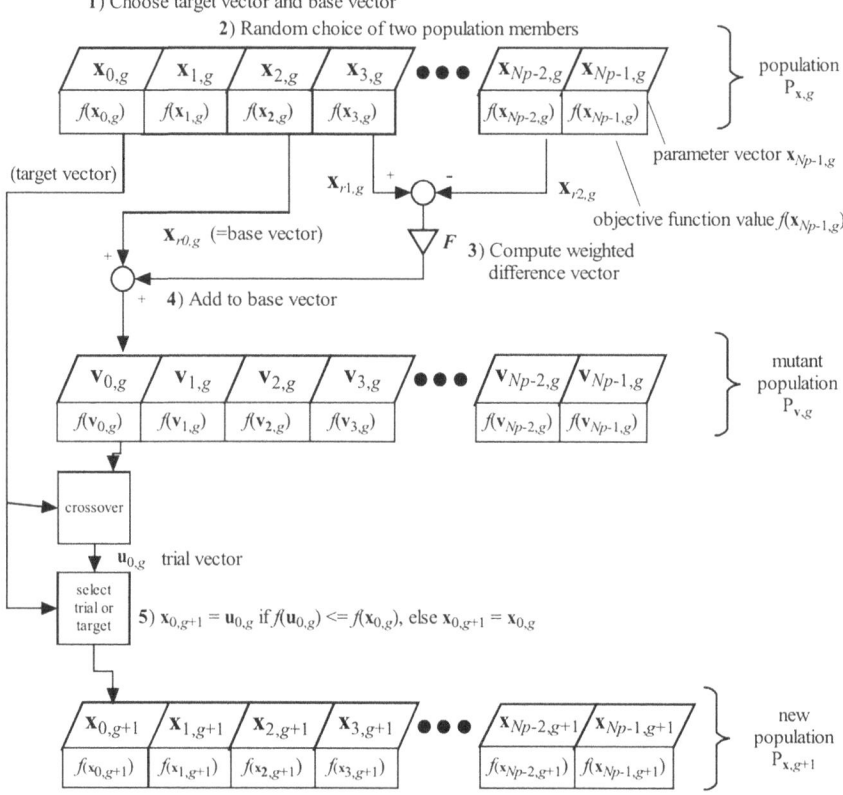

Fig. 2.6 A flowchart representation of the Diffirential Evolution optimization algorithm [1]

The combination of mutation and crossover allows DE to perform acceptably well on functions that are either decomposable[4] or non-decomposable, while maintaining rotational invariance[5] [1]. Figure 2.6 illustrates DE with a flowchart.

There are also a plethora of strategies to choose the three candidate vectors for the mutation, which can have dramatic effects on the convergence speed on some problems. For example, the base of the mutation (x_{r3} in the previous examples) could be the best candidate from the previous generation instead of a random one, to make the algorithm more greedy. The distribution that is used to check the crossover condition C_r can also be changed, as well as the number of vectors used to generate the mutation.

---

[4] A decomposable function can be written as the sum of D one-dimensional functions: $f(\mathbf{x}) = \sum_{i=1}^{D} f_i(x_i)$.

[5] An algorithm is rotationally invariant if its performances do not depend on the objective function being aligned with a privileged coordinate system.

In general, strategies are indicated with triplets, for example `DE/rand/1/bin` indicates the "standard" DE (which we just described), where a random vector is used as the base for the mutation against one difference of two other vectors, and crossover is checked against a binomial distribution; `DE/best/1/exp` instead is a more greedy variant that uses the best candidate from the previous generation as base for the mutation, and the crossover is checked using an exponential distribution.

To summarize, there are five basic parameters which define the behaviour of DE:

- Population size N_p
- Mutation scale F
- Crossover threshold C_r
- Selection strategy: `DE/rand/1/bin`, `DE/best/1/exp`, `DE/best/2/bin`, `DE/rand-to-best/1/bin`, ...
- Initial distribution of the N_p vectors in the allowed space: random, uniform grid, halton, sobol, ...

Recommendations and performance evaluations on many kind of problems can be found in [1, 41]. In general, populations strongly in excess of $2D$ vectors (where D is the number of free parameters, hence the size of the vectors) do not provide higher probability of convergence while rising the computational cost significatively instead. More detailed analysis of convergence properties can be found in [42].

Chapter 3
Examples of ODE Models of Physical Phenomena

This section presents and discusses several examples of using ODEs to model and study various physical phenomena. The examples are organized to show increasing complexity, beginning with single equations involving one variable and progressing to systems of equations with multiple variables. The examples below simplify the discussion by omitting units of measurement for the parameters used in the models. This approach helps keep the analysis clear and avoids complicating it with unit conversions.

3.1 Simulating Population Growth

In this section, an ODE model will be developed to simulate a social phenomenon: the dynamics of a population. Specifically, the model will be based on Malthus' exponential growth model (1798). The model assumes a homogeneous population, meaning all individuals exhibit identical behavior, and factors such as age, sex, social class, etc., are negligible. The population, denoted by the function $N(t)$, represents the number of individuals at time t. Moreover, the model assumes that the population resides in an isolated habitat, unaffected by immigration, emigration, or external factors altering living conditions. Under these conditions, the changes in population size over time are solely influenced by births and deaths.

According to Malthus' model, the rate of change of the population, $\frac{dN}{dt}$, can be expressed as:

$$\frac{dN}{dt} = \text{birth} \cdot N(t) - \text{death} \cdot N(t) = r \cdot N(t) \tag{3.1.1}$$

where $r = \text{birth} - \text{death}$ represents the net growth rate of the population. The initial population size at time $t = 0$ is given by $N(0) = 10$. This differential equation reflects the assumption that the population grows exponentially in the absence of limiting factors, with the growth rate determined by the difference between the birth and death rates.

```python
# Import necessary libraries
import numpy as np
import matplotlib.pyplot as plt
from scipy.integrate import solve_ivp

# Define the differential equation for the population
↪    growth model
def population_growth(t, n):
    # dn/dt = r * n, where r is the net growth rate
    ↪    (birth rate - death rate)
    return r * n

# Set the initial population size, growth rate, and time
↪    span for simulation
n0 = 10.0    # Initial population size at time t = 0
r = -0.2     # Net growth rate (negative value indicates
↪    population decline)
t_span = (0, 10)  # Time interval for the simulation
↪    (from t = 0 to t = 10 years)
t_eval = np.linspace(t_span[0], t_span[1], 100)   # Time
↪    points at which to store the computed solution

# Solve the ODE using solve_ivp
solution = solve_ivp(population_growth, t_span, [n0],
↪    t_eval=t_eval)

# Plot the results
plt.figure()
plt.plot(solution.t, solution.y[0], label='Population
↪    size over time')
plt.xlabel('Time (Years)')        # Label for the x-axis
plt.ylabel('Population size, n(t)')  # Label for the
↪    y-axis
plt.title('Population Dynamics Based on Malthusian
↪    Growth Model')   # Plot title
plt.grid(True)   # Add a grid to the plot for better
↪    readability
plt.legend()   # Add a legend to the plot
plt.show()   # Display the plot
```

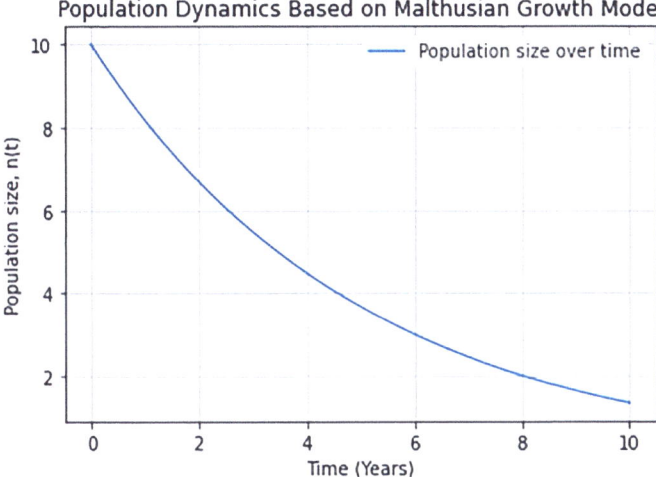

To simulate the effect of immigration and emigration in the population model, the differential equation can be modified to include additional terms representing these factors. Immigration increases the population, while emigration decreases it. The rates of immigration and emigration can be assumed to be constants, denoted as *immigration_rate* and *emigration_rate*, respectively. For example, to simulate the effect of constant immigration (5 individuals per unit time) and emigration (2 individuals per unit time) starting from the 6th year, the model must include these changes conditionally. This can be achieved by modifying the differential equation to incorporate the immigration and emigration rates only after a specific time threshold.

Below, the updated code with the additional terms for immigration and emigration.

```
1    # Import necessary libraries
2    import numpy as np
3    import matplotlib.pyplot as plt
4    from scipy.integrate import solve_ivp
5
6    # Define the differential equation for the population
     ↪   growth model with conditional immigration and
     ↪   emigration
7    def population_growth(t, n):
8        # Define immigration and emigration rates
         ↪   conditionally based on time
9        if t >= 6:
10           immigration_rate = 5.0   # Immigration rate
             ↪   (people per unit time) after year 6
11           emigration_rate = 2.0    # Emigration rate
             ↪   (people per unit time) after year 6
12       else:
13           immigration_rate = 0.0   # No immigration before
             ↪   year 6
14           emigration_rate = 0.0    # No emigration before
             ↪   year 6
```

```
15
16        # dn/dt = r * n + immigration_rate - emigration_rate
17        return r * n + immigration_rate - emigration_rate
18
19    # Set initial condition, rate of growth, and time span
20    n0 = 10.0    # Initial population size at time t = 0
21    r = -0.2     # Net growth rate (birth rate - death rate)
22    t_span = (0, 10)   # Time interval for the simulation
      ↪  (from t = 0 to t = 10 years)
23    t_eval = np.linspace(t_span[0], t_span[1], 100)   # Time
      ↪  points at which to store the computed solution
24
25    # Solve the ODE using solve_ivp
26    solution = solve_ivp(population_growth, t_span, [n0],
      ↪  t_eval=t_eval)
27
28    # Plot the results
29    plt.figure()
30    plt.plot(solution.t, solution.y[0], label='Population
      ↪  size over time')
31    plt.xlabel('Time (Years)')       # Label for the x-axis
32    plt.ylabel('Population size, n(t)')  # Label for the
      ↪  y-axis
33    plt.title('Population Dynamics with Conditional
      ↪  Immigration and Emigration')  # Plot title
34    plt.grid(True)   # Add a grid to the plot for better
      ↪  readability
35    plt.legend()   # Add a legend to the plot
36    plt.show()   # Display the plot
```

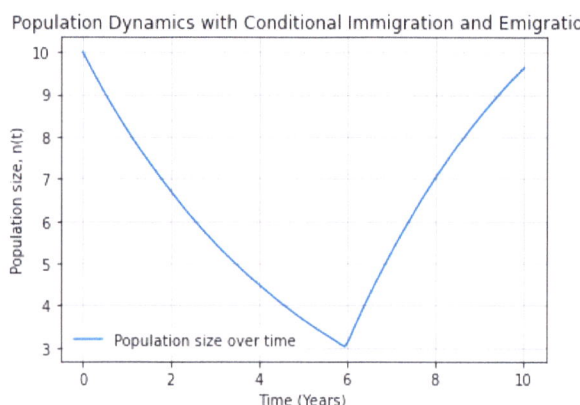

This extended model is more realistic as it accounts for individuals entering and leaving the population, which are common factors in real-world population dynamics.

3.2   Balance Equation

In the context of population dynamics, the Malthus' equation can be seen as a specific case of a broader class of *balance equations* used in various fields. In engineering, there are four common balance equations derived from conservation principles: mass, momentum, energy, and species. Each of these balance equations follows a general form, which can be expressed as:

$$\text{Accumulation} = \text{Inlet} - \text{Outlet} + \text{Generation} - \text{Consumption} \qquad (3.2.2)$$

Here, the accumulation term on the left side represents the rate of change of a property (such as mass, energy, etc.) within a defined system or control volume. On the right side, the inlet term accounts for the flow of the property into the system, the outlet term represents the flow out of the system, the generation term denotes any creation of the property within the system, and the consumption term indicates any consumption or loss of the property.

For the Malthus model, the equation can be adapted to fit this general form as follows:

$$\frac{dN}{dt} = \text{Inlet} - \text{Outlet} + \text{Generation} - \text{Consumption} \qquad (3.2.3)$$

In the simplest Malthus model, the 'Inlet' and 'Outlet' terms are combined into the net growth rate $r \cdot N$, with no explicit generation or consumption terms. However, when extending the model to include immigration and emigration, these terms can be incorporated into the equation to reflect additional factors affecting population size. Thus, the Malthus model exemplifies how differential equations can be used to represent balance equations in various domains, providing insight into the dynamics of systems governed by conservation principles.

Thus, to build and solve ODEs, two essential components are required: the balance equation and the initial condition. The balance equation describes the fundamental relationships and interactions governing the system, while the initial condition provides the starting point for the simulation. Together, they enable the accurate modeling and analysis of dynamic systems across various applications. The subsequent examples will illustrate models of increasingly complex balance equations, demonstrating how additional factors and interactions are accounted for as the complexity of the system increases.

3.3   Simulating Prey-Predator Dynamics

The prey-predator model extends the concept of the Malthus model by incorporating interactions between two species: prey and predators. The dynamics are described by the following ODEs system:

$$\begin{cases} \frac{dN_{\text{prey}}}{dt} = \alpha N_{\text{prey}} - \beta N_{\text{prey}} N_{\text{predator}} \\ \frac{dN_{\text{predator}}}{dt} = \delta N_{\text{prey}} N_{\text{predator}} - \gamma N_{\text{predator}} \end{cases} \qquad (3.3.4)$$

where:

- N_{prey} is the number of prey,
- N_{predator} is the number of predators,
- α is the prey birth rate,
- β is the rate at which predators consume prey,
- δ is the rate at which consumed prey increase the predator population,
- γ is the natural death rate of predators.

This model introduces interactions between the two species, unlike the Malthus model, which considers a single population growing exponentially. Below is the Python code to solve and analyze the prey-predator model using the `solve_ivp` function from the `scipy.integrate` library.

```python
# Import necessary libraries
import numpy as np
import matplotlib.pyplot as plt
from scipy.integrate import solve_ivp

# Define the differential equations for the
#     prey-predator model
def prey_predator(t, y):
    N_prey, N_predator = y
    alpha = 0.1   # Prey birth rate
    beta = 0.02   # Predation rate
    delta = 0.01   # Predator reproduction rate (from
    #     consumed prey)
    gamma = 0.1   # Predator natural death rate

    # Define the differential equations
    dN_prey_dt = alpha * N_prey - beta * N_prey *
    #     N_predator
    dN_predator_dt = delta * N_prey * N_predator - gamma
    #     * N_predator

    return [dN_prey_dt, dN_predator_dt]

# Set initial conditions and time span
N_prey0 = 40.0   # Initial prey population
N_predator0 = 9.0   # Initial predator population
initial_conditions = [N_prey0, N_predator0]
t_span = (0, 50)   # Time span for the simulation
t_eval = np.linspace(t_span[0], t_span[1], 500)   # Time
#     points for evaluation

# Solve the ODE using solve_ivp
solution = solve_ivp(prey_predator, t_span,
    initial_conditions, t_eval=t_eval)
```

```
29
30   # Plot the results
31   plt.figure(figsize=(8, 4))
32
33   # Plot prey and predator populations
34   plt.subplot(1, 2, 1)
35   plt.plot(solution.t, solution.y[0], label='Prey
     ↪   Population')
36   plt.plot(solution.t, solution.y[1], label='Predator
     ↪   Population')
37   plt.xlabel('Time')
38   plt.ylabel('Population')
39   plt.title('Prey-Predator Dynamics')
40   plt.legend()
41   plt.grid(True)
42
43   # Phase plot (Prey vs. Predator)
44   plt.subplot(1, 2, 2)
45   plt.scatter(solution.y[0], solution.y[1], marker='.',
46                 c=plt.colormaps['viridis'].resampled(len(sol
                  ↪   ution.y[0])).colors)
47   plt.xlabel('Prey Population')
48   plt.ylabel('Predator Population')
49   plt.title('Phase Plot')
50   plt.grid(True)
51
52   plt.tight_layout()
53   plt.show()
```

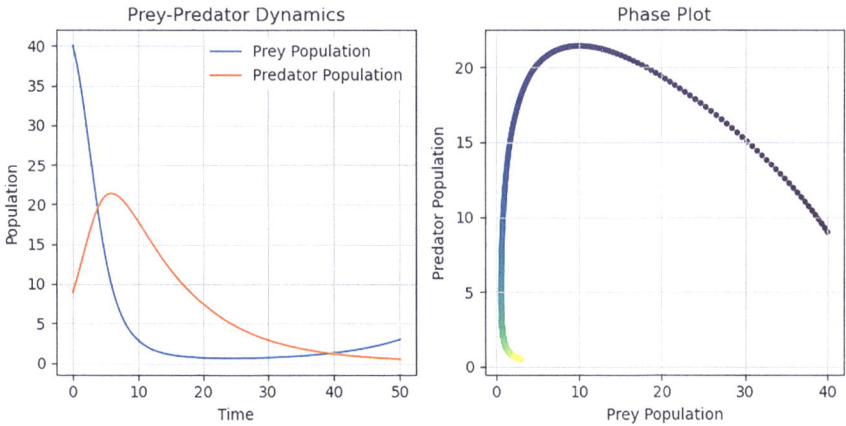

The prey-predator model involves a system of coupled ODEs reflecting the interactions between two species. This contrasts with the Malthus model, which describes the exponential growth of a single population without interactions. The parameters

(α, β, δ, γ) represent rates that describe the interactions between prey and predators, such as the prey birth rate, the consumption rate by predators, the increase in predator population from consumed prey, and the natural death rate of predators. The `solve_ivp` function is used to integrate the system of ODEs over time. The results are visualized in two plots: one showing the populations of prey and predators over time and another showing the phase plot, which illustrates how prey and predator populations interact.

In particular, the phase plot, also known as a phase portrait, is a graphical representation that illustrates the trajectories of a dynamical system in phase space. For the prey-predator model, the phase plot visualizes the relationship between the prey and predator populations over time, providing insights into the cyclical dynamics and stability of the system. The X-Axis represents the prey population N_{prey}, whereas the Y-Axis (Vertical) represents the predator population $N_{predator}$.

The plot traces the path that the system follows over time in the prey-predator population space. Each point on the trajectory corresponds to the state of the system at a particular time, with specific prey and predator population values. In many prey-predator systems, especially those described by the Lotka-Volterra equations (which the given model resembles), the populations exhibit cyclical behavior. This means that the populations of prey and predators rise and fall in a regular pattern. The phase plot often reveals closed or nearly closed loops, indicating these cycles. The system moves around these loops in a specific direction (either clockwise or counterclockwise), showing how the populations change relative to each other over time. When the system is inside the loop, it indicates lower amplitudes in population fluctuations. Points on the loop represent specific states where the populations reach certain combinations that repeat over time. Arrows can be added to the trajectory to indicate the direction of time progression, helping to understand how the system evolves.

The phase plot can also highlight equilibrium points where the populations remain constant. For the prey-predator model, a common equilibrium occurs when the growth of prey is exactly balanced by predation, and predator deaths are balanced by the reproduction rate due to consuming prey. At these points, the trajectory would ideally settle, indicating a stable system. However, in the basic Lotka-Volterra model without additional factors (like carrying capacity or more complex interactions), the equilibrium is often neutrally stable, leading to perpetual cycles.

Changing parameters such as the prey birth rate (α), predation rate (β), predator reproduction rate (δ), and predator death rate (γ) alters the shape and size of the trajectories in the phase plot. For instance, increasing the predator's death rate might lead to smaller predator populations, shifting the trajectory accordingly.

While the Malthus model focuses on the exponential growth of a single population, the prey-predator model extends this concept by including interactions between two populations. This results in a more complex system where the dynamics of one population depend on the interactions with the other. The prey-predator model demonstrates how differential equations can be used to represent more complex interactions in ecological systems, building on the foundational concepts introduced by the Malthus model.

3.4 Diabetes Mellitus

Characterized by sustained high blood sugar levels, diabetes is most commonly due to the pancreas not being able to produce enough insulin, or to body cells becoming unresponsive to the hormone effects. A relatively simple model, that considers only the interaction between glucose and insulin, can be use for the diagnosis of the condition, as initially presented in [43] and recently expanded on in [44].

Let $G(t)$, $I(t)$ be the Glucose and Insulin level over time. The system (autonomous and hence independent from the absolute time) is defined as follows:

$$G'(t) = F_G(G(t), I(t)) \qquad (3.4.5)$$
$$I'(t) = F_I(G(t), I(t)) \qquad (3.4.6)$$

and is assumed to be such that there exist an optimal steady steate (or equilibrium point) $\bar{G}$, $\bar{I}$ such that:

$$F_G(\bar{G}, \bar{H}) = F_I(\bar{G}, \bar{I}) = 0 \qquad (3.4.7)$$

Since we expect a subject to always go back to his or hers optimal state $\bar{G}$, $\bar{I}$ after the system is pertubed (by a meal, for example), this must be an asymptotically stable equilibrium point of the system. The equilibrium point can be translated to be in $(0, 0)$ by using the translated variables:

$$g(t) = G(t) - \bar{G}, \quad i(t) = I(t) - \bar{I} \qquad (3.4.8)$$

and the system therefore becomes:

$$g'(t) = F_g(g(t), i(t)) \equiv F_G(g(t) + \bar{G}, i(t) + \bar{I}) \qquad (3.4.9)$$
$$i'(t) = F_i(g(t), i(t)) \equiv F_I(g(t) + \bar{G}, i(t) + \bar{I}) \qquad (3.4.10)$$

Assuming F_g, F_i are sufficiently regular, they can be expanded using Taylor's serie to obtain the linearized system:

$$g' = m_{gg}g + m_{gi}i + \gamma_g(g, i) \qquad (3.4.11)$$
$$i' = m_{ig}g + m_{ii}i + \gamma_i(g, i) \qquad (3.4.12)$$

Remembering that a zero value for g and i in this transposed system means that glucose and insuline are at their respective optimal levels, it is possible to reason on the sign of the four m parameters:

- as g must decrease when i increases, m_{gi} must be negative;
- i must increase when g increases, hence m_{ig} must be positive;
- in absense of glucose, insulin tends to return to zero, hence m_{ii} must be negative;

- glucose is in any case slowly used by the organism even in absence of insulin, hence also m_{gg} must be negative (and probably "small").

We can therefore represent the system as:

$$\begin{pmatrix} g' \\ i' \end{pmatrix} = \begin{pmatrix} -m_{gg} & -m_{gi} \\ m_{ig} & -m_{ii} \end{pmatrix} \begin{pmatrix} g \\ i \end{pmatrix} + \begin{pmatrix} \gamma_g(g, i) \\ \gamma_i(g, i) \end{pmatrix} \qquad (3.4.13)$$

or equivalently:

$$\mathbf{y}'(t) = A\mathbf{y}(t) + \mathbf{g}(\mathbf{y}(t)) \qquad (3.4.14)$$

where we assume the signes are given explicitly in the representation, and hence the four m parameters are all positive. In this case, the characteristic polynomial of A is given by:

$$p(\lambda) = \det(A - \lambda I) = \lambda^2 + (m_{gg} + m_{ii})\lambda + (m_{gg}m_{ii} + m_{gi}m_{ig}). \qquad (3.4.15)$$

Since this is a quadratic equation, the following relationships hold[1]:

$$\lambda_1 + \lambda_2 = -(m_{gg} + m_{ii}) < 0, \quad \lambda_1\lambda_2 = m_{gg}m_{ii} + m_{gi}m_{ig} > 0. \qquad (3.4.16)$$

This implies that the eigenvalues λ_1 and λ_2 must have negative real parts, which in turn indicates the asymptotic stability of the linear part of system (3.4.14).

Furthermore, since the nonlinear components γ_g and γ_i were derived as the remainder of Taylor's expansion from the second term, it follows that:

$$\|\mathbf{g}(\mathbf{y}(t))\| = O(\|\mathbf{y}(t)\|^2) \Rightarrow \lim_{y \to 0} \frac{\|\mathbf{g}(\mathbf{y})\|}{\|\mathbf{y}\|} = 0. \qquad (3.4.17)$$

Thus, the conditions for uniform asymptotic stability are always satisfied. It does therefore make sense to ignore the non-linear part (which are actually unknow, as the actual functions that represent this system are unknown!) and use the simplified linearized system to still make meaningful predictions on the asymptotic behaviour of the system $\mathbf{y}' = A\mathbf{y}$.

It is easier to ignore the units of measure for the moment, and consider g, i and t to be indicated in some standard unit of glucose, insulin and time. First of all, let's define the simulation's temporal extremes, starting point, and a function generator that will allow to simulate different conditinos. In particular, we can still add a know term which we control to simulate external stimulation. The system becomes:

$$\mathbf{y}'(t) = A\mathbf{y}(t) + \mathbf{g}(t) \qquad (3.4.18)$$

---

[1] In any second-degree equation $ax^2 + bx + c$, the roots satisfy $x_1 + x_2 = -\frac{b}{a}$ and $x_1 x_2 = \frac{c}{a}$.

where

$$
\mathbf{g}(t) = \begin{cases} \begin{pmatrix} k \\ 0 \end{pmatrix} & \text{if } t_{g-in} < t < t_{g-out} \\ \mathbf{0} & \text{otherwise} \end{cases}
\tag{3.4.19}
$$

which allows to simulate a constant injection of k units of glucose starting at t_{g-in} and ending at t_{g-out}.

```python
import numpy as np
from matplotlib import pyplot as plt
from scipy.integrate import solve_ivp

t_min = 0; t_max = 100
k = 5.; t_g_in = 10; t_g_out = 20
y0 = np.array([0,0])

def g(t):
    if t > t_g_in and t < t_g_out:
        return np.array([k,0.0])
    else:
        return np.array([0,0])

def f_gen(A):
    return lambda t, y: A.dot(y) + g(t)

def solve_and_plot(f):
    res = solve_ivp(f, [t_min,t_max], y0,
                    max_step=1.,
                    method="BDF")
    y = res['y']
    t = res['t']

    fig, (ax1, ax2) = plt.subplots(1,2)
    fig.set_size_inches((8,4))
    ax1.plot(t, y[0], label='g')
    ax1.plot(t, y[1], label='i')
    ax1.set_xlabel('t')
    ax1.grid()
    ax1.legend()
    ax2.scatter(y[0], y[1], marker='.',
                c=plt.colormaps['viridis'].resampled(len
                ↪   (y[0])).colors)
    ax2.set_xlabel('g')
    ax2.set_ylabel('i', rotation='horizontal')
    ax2.grid()
```

Let's first try to look at what a healthy system might look like. Let's hypotesize that the interactions between insulin and glucose have the same magnitude, as well as insulin reabsorption rate in absence of stimuli; we therefore arbitrarily set $m_{gi} =$

$m_{ig} = m_{ii} = 0.5$. Glucose is expected to be slowly used by the body in general, hence it makes sense to set $m_g g = 0.01$, a relatively smaller value.

```
1    m_gg = -0.01; m_gi = -0.5
2    m_ig = 0.5; m_ii = -0.5
3    A = np.array([[m_gg, m_gi],[m_ig, m_ii]])
4    solve_and_plot(f_gen(A))
```

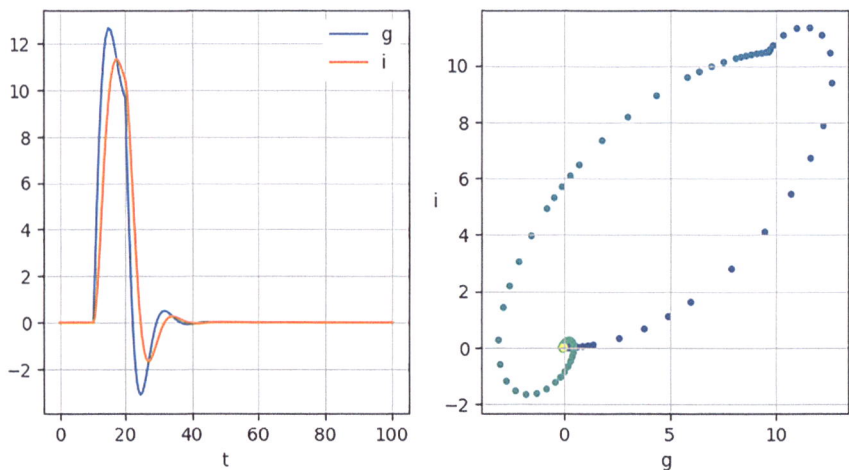

The system does indeed behave like one would expect (and as is verified in clinical trials): glucose and insulin levels are stable in absence of stimuli. As the glucose starts to spike (thanks to the external stimuli simulated by $\mathbf{g}(t)$ to start at t_{g-in}, insulin rises to tame the glucose spike. After the stimulus ends, levels quickly return to the asymptotic stable null solution after a few small oscillations.\ The glucose spike was successfully tamed by the insulin, and quickly returned to it's optimal level.

What happens if instead, for example, the pancreas is not able to produce enough insulin (Type I diabetes)? In our model, that would be represent by a smaller value of m_{ig}:

```
1    m_gg = -0.01; m_gi = -0.5
2    m_ig = 0.05; m_ii = -0.5
3    A = np.array([[m_gg, m_gi],[m_ig, m_ii]])
4    solve_and_plot(f_gen(A))
```

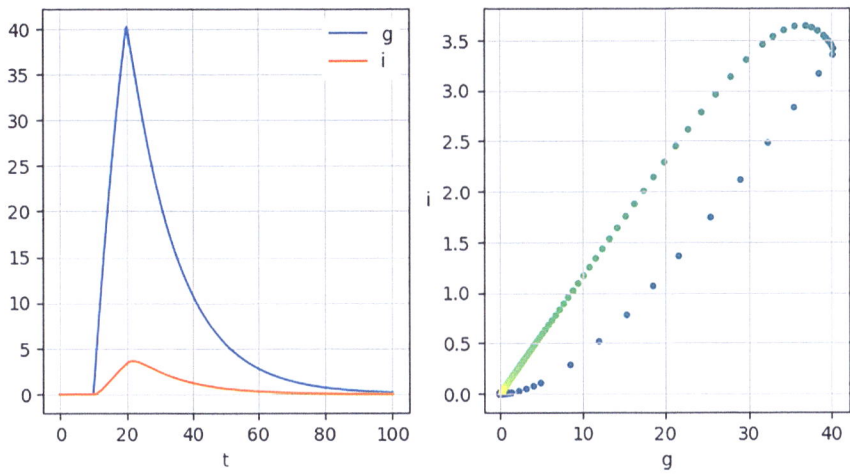

The resulting glucose spike is much stronger, and the glucose takes a really long time to go back to its normal value. Similarly, if there is bad insulin sensitivity (small m_{gi}), insulin is produced but the body is not reacting to it, result in a huge insuline spike which does, however, not have a significant effect on the glucose:

```
1   m_gg = -0.01; m_gi = -0.05
2   m_ig = 0.5; m_ii = -0.5
3   A = np.array([[m_gg, m_gi],[m_ig, m_ii]])
4   solve_and_plot(f_gen(A))
```

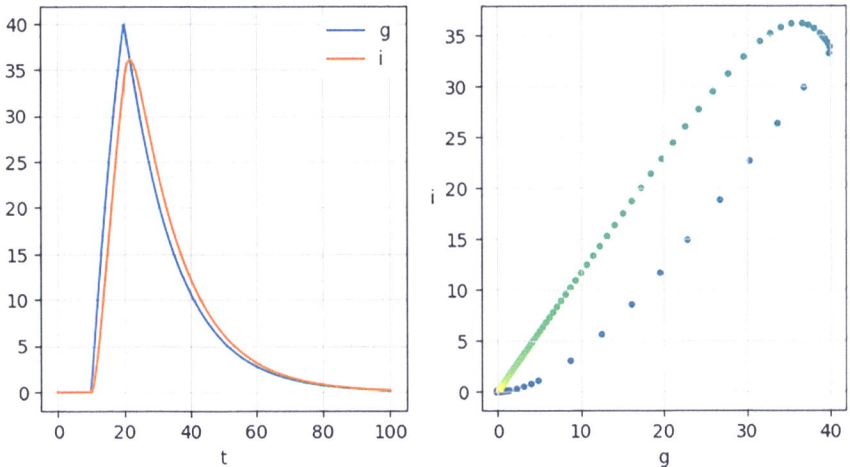

In Type II diabetes, these two effects combine, leading to a higher and longer glucose spike:

```
1    m_gg = -0.01; m_gi = -0.1
2    m_ig = 0.1; m_ii = -0.5
3    A = np.array([[m_gg, m_gi],[m_ig, m_ii]])
4    solve_and_plot(f_gen(A))
```

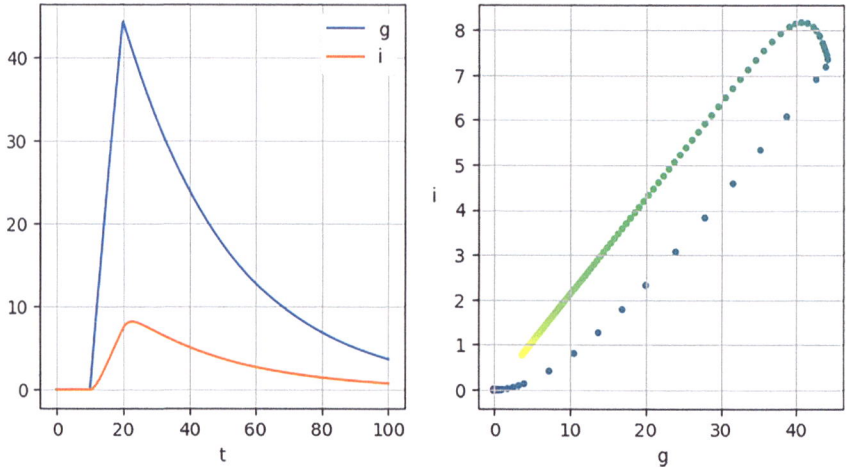

Finally, suppose that the pancreas is producing insulin normally and the body reacts accordingly, but the insulin regulation process does not work well (m_{ii} is small). In this case, we could expect the patient to experience wild glucose spikes at regular intervals, which can prove to be quite dangerous:

```
1    m_gg = -0.01; m_gi = -0.5
2    m_ig = 0.5; m_ii = -0.05
3    A = np.array([[m_gg, m_gi],[m_ig, m_ii]])
4    solve_and_plot(f_gen(A))
```

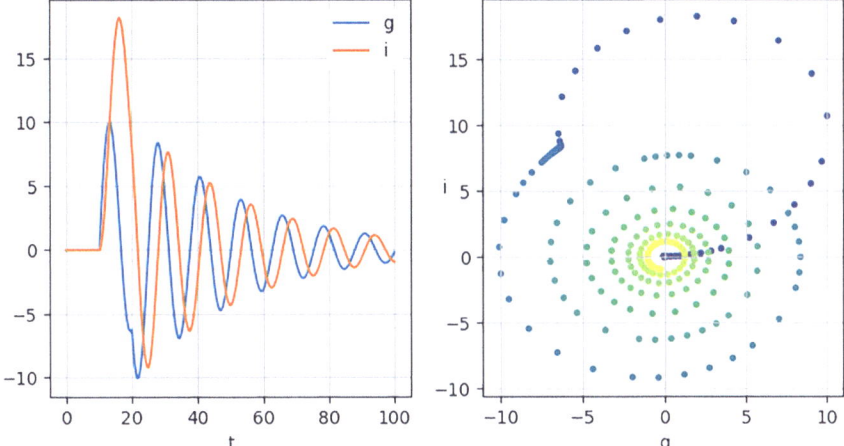

Of course, this extremely simplified model has sever limitations, but can nevertheless be useful.

Let's now look at how this simple model can be fitted on real data. In [45], the authors perform a standardized glucose reaction test by administering a fixed dose of glucose and measuring for three hours, every 30 min, the blood levels of many important indicators, including glucose and insulin levels.

The data is provided in tabular form, with the corresponding glucose and insuline blood provided at times $t = 0, 30, 60, 90, 120, 150, 180$. We will consider as free parameters of our model only the matrix A, hence the values $m_g g, m_g i, m_i g, m_i i$. First of all, it is necessary to define an error measure that allows to use an optimization method to find the model parameters. In this case, we want to measure the distance between the glucose and insulin curve predicted by the model, and the ones provded by the data.

The experimental measurements provide only a few points, which in some cases may not be enough to direct the optimizer in right direction. It is possible to expand the available points by means of interpolation. Since we expect glucose and insulin levels to change smoothly, a good choice could be a cubic sipline interpolation. This can be easily achieved using scipy's CubicSpline class as follows:

```
from scipy.interpolate import CubicSpline
T = np.array([0, 30, 60, 90, 120, 150, 180])
Y_g = np.array([4.8, 8, 7.9, 6.7, 5.8, 4.6, 3.8])
x = np.linspace(T[0], T[-1], 50)
points = CubicSpline(T, Y_g, bc_type='clamped')(x)
plt.plot(x, points)
plt.plot(T,Y_g, 'o')
plt.xlabel('t')
plt.ylabel('y', rotation='horizontal')
plt.grid()
```

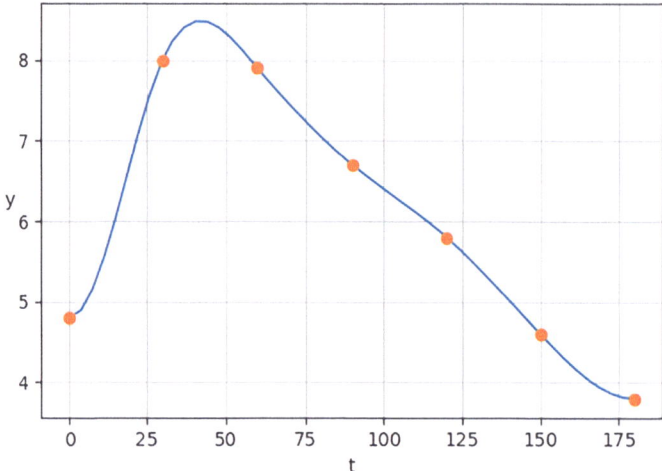

In the above graph, the orange dots mark the available data, and the continuous line the cubic spline interpolation. Since the interpolation is a continuous function, we can sample it at any point. After arbitrarily deciding that it is best to use 100 points to check the model's fitness, we can, in this simple example, compute it simply as the sum of the squared distances betwee the model's result and the interpolation function sampled at the same point time. Let d_g, d_i be the interpolating functions for glucose and insuline data respectively, the error can then be computed as

$$e = \sum_{x \in T}(d_g(x) - y_g(x))^2 + 0.2 \sum_{x \in T}(d_i(x) - y_i(x))^2$$

where y_g, y_i are the model's result, and $T = \{t_0, ..., t_n\}$ is the list of time ticks indi-viduated by the solver. For the sake of theis example, we also decided that the insulin data is less precise, and therefore the model's errors on insulin are scaled down in the error function to give them less importance during the optimization phase. We are therefore going to perform a sort of multivariate least squares approximation to choose the model's parameters that produce a simulation as near as possible to the provided data.

Since our model expects the base glucose and insuline level to be zero, we translate all the provided data by substricting the first value, which is supposed to be the baseline for both glucose and insuline. Furthermore, insuline levels will be scaled down by a factor of ten: this does not affect the behavior of the system, but brings the insuline values in the same range as the glucose ones, allowing for easier to read plots.

To simulate the initial glucose spike, assuming a linear absorpion for simplicity, we can use the same $\mathbf{g}$ as previously defined in (3.4.18); since we expect the peak to happen at the 30 min mark with magnitude given by the respective measurement m, the constant k will have to be $k = m/30$.

It is now time to define the error function:

```
def g(t, k):
    if t > 0 and t < 30:
        return np.array([k,0.0])
    else:
        return np.array([0,0])

def f_gen(A, k):
    return lambda t, y: A.dot(y) + g(t, k)

def error_function_generator(T, Y_g, Y_i, y0):
    k = Y_g[1] / (t_g_out-t_g_in)
    def error_function(v):
        A = np.array([[v[0], v[1]],[v[2], v[3]]])
        res = solve_ivp(f_gen(A, k), [T[0],T[-1]], y0,
                        method="BDF",
                        t_eval=np.linspace(T[0], T[-1], 100),
                        #t_eval=T[2:]
                        )

        ref_g = CubicSpline(T, Y_g,
          ↪   bc_type='clamped')(res['t'])
        error = np.sum((res['y'][0]-ref_g)**2)

        ref_i = CubicSpline(T, Y_i,
          ↪   bc_type='clamped')(res['t'])
        error += (np.sum((res['y'][1]-ref_i)**2))/2.

        return error

    return error_function

T = np.array([0, 30, 60, 90, 120, 150, 180])
t_g_in, t_g_out = 0, 30
```

err_f_gen is a generator that given as parameters the data time marks **T**, the tabular data for glucose and insulin $\mathbf{Y_g}, \mathbf{Y_i}$ and the initial conditions $\mathbf{y_0}$ (which in this case, are always $(0, 0)$ by construction), returns an error function that accepts as parameter the vector of values that compose the matrix A and that the optimizer must find. We can finally define the optimizer function, and a helper for plotting:

```python
from scipy.optimize import minimize, Bounds
v_init = [-0.01,-0.01,0.01,-0.01]

def optimize(T, Y_g, Y_i):
    error_function = error_function_generator(T, Y_g,
    ↪  Y_i, [Y_g[0], Y_i[0]])
    res = minimize(error_function, v_init,
    ↪  method='nelder-mead')
    A = np.array([[res.x[0], res.x[1]],[res.x[2],
    ↪  res.x[3]]])
    k = Y_g[1] / (t_g_out-t_g_in)
    return A, k

def solve_and_plot(f, T, Y_g, Y_i, y0):
    res = solve_ivp(f, [T[0],T[-1]], y0,
                        method="BDF")
    y = res['y']
    t = res['t']
    ref_g = CubicSpline(T, Y_g, bc_type='clamped')(t)
    ref_i = CubicSpline(T, Y_i, bc_type='clamped')(t)
    fig, (ax1, ax2) = plt.subplots(1,2)
    fig.set_size_inches((8,4))

    ax1.plot(t, ref_g, '--', label='g ref.', c='green')
    ax1.plot(T, Y_g, 'o', c='green')

    ax1.plot(t, ref_i, '-.', label='i ref.', c='grey')
    ax1.plot(T, Y_i, 'x', c='grey')

    ax1.plot(t, y[0], label='g', c='blue')
    ax1.plot(t, y[1], label='i', c='orange')
    ax1.set_xlabel('t')
    ax1.grid()
    ax1.legend()
    ax2.scatter(y[0], y[1], marker='.',
                c=plt.colormaps['viridis'].resampled(len
                ↪  (y[0])).colors)
    ax2.set_xlabel('g')
    ax2.set_ylabel('i', rotation='horizontal')
    ax2.grid()
```

The minimize function in this case will use the Nelder-Mead method to search the parameters space for a combination that globally minimizes the error function. This is one of the simplest minimazion algorithms, but is usually worth trying before moving to more complex approaches. Now everything is in place to try and fit the model on real data. This first set of data correspond to the average monophasic case described in [45]:

```
1   Y_g = np.array([4.8, 8, 7.9, 6.7, 5.8, 4.6, 3.8]) - 4.8
2   Y_i = (np.array([6.4, 45.8, 56.6, 47.7, 31.4, 15.5,
    ↪   6.8])-6.4) / 10.
3   A, k = optimize(T, Y_g, Y_i)
4   #A, k = np.array([[-0.05, -0.5],[0.1, -0.01]]), 0.32
5   solve_and_plot(f_gen(A, k), T, Y_g, Y_i, [Y_g[0],
    ↪   Y_i[0]])
```

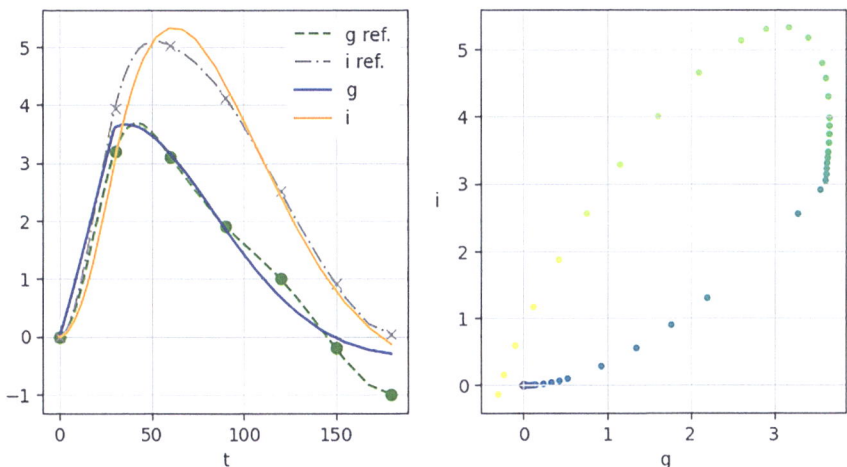

The seemilngly linear glucose growth in between 0 and 30 minutes is caused by the external stimulus function **g**; there is however already also a contribution from the model in this time interval, so the function value at time 30 is not exactly the expected value *k* but may differ.

This simple model is evidently not able to exactly fit all the datapoints, but surely captures significantly well the behaviour of both glucose and insuline. Once such a model is fitted on a particular patient's data, it could be used for example to predict glucose levels given only one or two measurements instead of much higher frequencies. Let's see what happens in the biphasic case:

```
1   Y_g = np.array([4.7, 7.6, 6.6, 5.5, 4.3, 3.3, 3.9]) - 4.7
2   Y_i = (np.array([6., 50.2, 55.2, 38.1, 19.1, 6.9, 5.1])
    ↪   - 6.) / 10.
3   A, k = optimize(T, Y_g, Y_i)
4   solve_and_plot(f_gen(A, k), T, Y_g, Y_i, [Y_g[0],
    ↪   Y_i[0]])
```

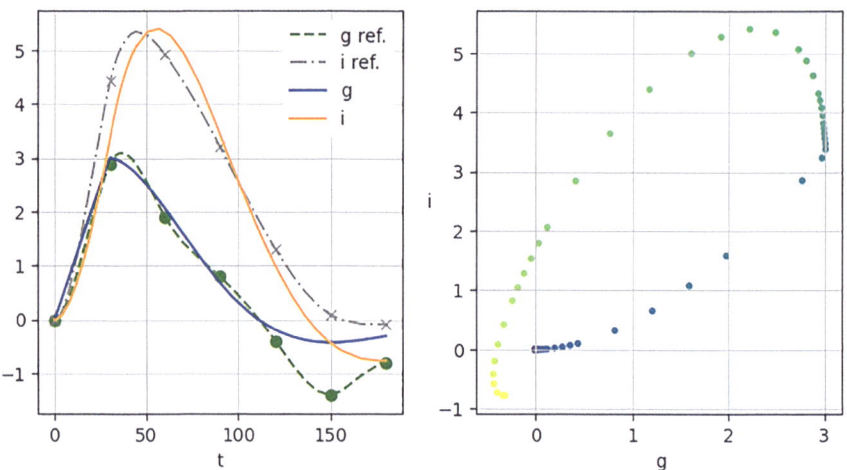

And the triphasic case:

```
1   Y_g = np.array([4.7, 7.2, 6.2, 5.1, 5.6, 4.9, 4]) - 4.7
2   Y_i = (np.array([5.9, 49, 44.9, 33, 29.4, 17.2, 6.4]) -
    ↪   5.9) / 10.
3   A, k = optimize(T, Y_g, Y_i)
4   solve_and_plot(f_gen(A, k), T, Y_g, Y_i, [Y_g[0],
    ↪   Y_i[0]])
```

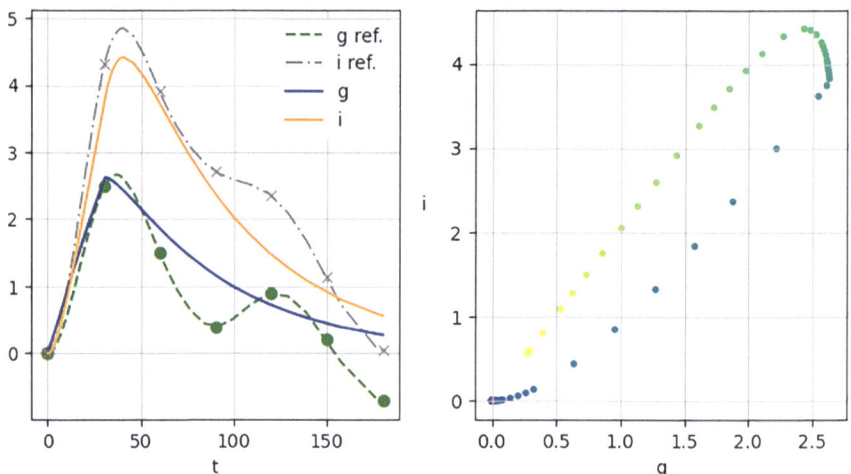

The triphasic case doesn't seem to be well fitted: all the higher order behaviors are ignored. This is most probably be due the limitations of the model which lacks all the non-linear components; as we have seen before, the model can easily produce oscillations around the origin, but the data we are trying to fit is strongly skewed in the positive values. This is the point where it make make sense to start tweaking the model; hovewer, with great complexity come great responsibility: the simplest model that is able to reproduce the fundamental aspects of the phenomenon one is trying to simulate sometimes may prove to be not only much faster and easier to develop, but also more useful for the intuitive understanding of the phenomenon being modelled. As for example with this example, we were able to predict that in some conditions one can expect multiple oscillations, albeit it may not be possible to quantify their magnitude by predicting real data, this is already a pretty significant insight in the behavior of the system!

Chapter 4
A Four-Step Method for Constructing System-Level Brain Models with ODE and Python

This chapter primarily describes a system-level approach to modeling the brain, focusing on the interactions between different brain areas and the overall neural activity trends of entire regions rather than individual areas or single neurons. The emphasis is on the dynamics of neural activity across networks of areas. In particular, the chapter covers the following topics. The first part introduces critical theoretical concepts necessary to understand the role of system-level models in simulating some brain functions. It explores the potential of using computational models to create 'digital twins' of subjects, simulating the effects of therapies before testing on actual subjects, and reducing future reliance on animal models. The second part details the method proposed in this book for constructing system-level models using Python and ODEs and includes several code examples.

4.1 Levels of Abstraction for Simulating Brain Functions

It is feasible to develop computational models to emulate brain operations by adopting various levels of abstraction. These models can range from detailed reproduction of microscopic phenomena occurring at the molecular and neuronal scale to broader models concentrating on mesoscopic and macroscopic phenomena [46]. For instance, some models delve into the minutiae of synaptic dynamics, regulating interactions within neuronal networks, while others zoom out to explore interactions among different brain regions. In general, the superiority of a model-whether highly detailed at the molecular level or more focused on mesoscopic or macroscopic aspects-is not assured. The choice of abstraction level depends on the specific problem at hand. For example, a detailed model of single-neuron functionality can shed light on the generation and propagation of action potentials, synaptic transmission regulation, and long-term synaptic plasticity and depression. However, certain phenomena may prove challenging to dissect solely through detailed computational models of single neurons. Large-scale network dynamics like brain waves and coordination between distant brain areas may elude analysis with single-neuron models. Furthermore, while detailed neuronal models can address aspects of synaptic plasticity, a comprehensive understanding of long-term memory processes necessitates higher-level considerations involving complex neural networks, feedback from subcortical structures, and interactions with the environment and the rest of the body. Single-neuron models often fall short of capturing the full spectrum of complex behaviors of living organisms. To scrutinize complex behavior, it becomes imperative to account for interactions between various neuron types, their connectivity, and modulation by hormones and other factors. Many neuropsychiatric disorders, such as major depression, bipolar disorder, schizophrenia, and neurodegenerative disorders like Parkinson's disease or Alzheimer's disease, involve a multifaceted interplay of genetic, environmental, and neural factors [47–49]. In all these instances, computational models at the mesoscopic and macroscopic levels may prove more adept at addressing the complexity of brain phenomena.

System-level models can replicate brain functions at both mesoscopic and macroscopic levels, offering several advantages [50–52]. By emphasizing interactions between different brain areas, these models provide a comprehensive understanding of overall brain functionality and the emergent properties resulting from these interactions. They enable the study of large-scale neural dynamics, revealing patterns and trends that smaller-scale examinations might miss. Furthermore, this approach proves more efficient and scalable, making it well-suited for analyzing complex neural networks. However, this perspective also has limitations. Focusing on broader interactions and trends may cause us to overlook specific details and mechanisms at the level of particular brain areas. This could lead to a loss of resolution and an incomplete capture of neural activity nuances. Additionally, the system-level approach might be less effective for examining the complexities of learning mechanisms, which often necessitate a more granular analysis.

Aside from computational modeling approaches at different levels of abstraction, scholars increasingly use machine learning tools to model brain functions, leveraging large datasets to predict and analyze neural behavior. Furthermore, statistical approaches offer valuable tools for understanding brain dynamics through the lens of probabilistic modeling. These approaches focus on inferring underlying neuronal processes from observed data, often using techniques such as Bayesian inference to uncover the hidden structure of neural activity and its relation to external stimuli or behavior. Overall, while the system-level approach offers a powerful framework for understanding large-scale neural dynamics, it is important to be aware of its limitations and complement it with other methods when necessary. Each modeling approach has strengths and weaknesses, and combining various methods according to the specific problem can offer a more comprehensive understanding of brain function.

4.2 Digital Twins

Digital twins represent virtual counterparts of objects, locations, individuals, or processes. For over two decades, they have found utility in the industrial sector by mimicking, akin to a video game, the functionality of tangible entities, assessing their performance, and delineating their constraints. This advancement has significantly contributed to the advancement of companies across diverse domains. In the manufacturing sector, for instance, digital twins facilitate the simulation and refinement of production processes, empowering companies to bolster efficiency, curtail expenses, and preemptively address potential issues before they manifest in production. Aerospace enterprises harness digital twins to virtually blueprint and evaluate aircraft prototypes, meticulously simulating their functionality across varied scenarios and conditions, thus streamlining the temporal and financial investments associated with tangible prototype development. Within the automotive domain, digital twins are instrumental in vehicle design optimization, crash test simulations, and the evaluation of autonomous driving systems. Beyond industrial applications, digital twins extend their utility to modeling and simulating ecosystem dynamics, climate variations, water resource management, and sustainable urban development [53, 54].

The creation and representation of digital twins can vary depending on the complexity of the systems they simulate. In engineering, investigators often develop digital twins by using mathematical models that employ equations to express the behavior of the physical system analytically. These models can subsequently be translated into computer programs, enabling simulation of their behavior. Engineers in manufacturing and construction companies commonly depict digital twins using 3D and CAD models. A CAD model, short for computer-aided design, serves as a digital portrayal of a three-dimensional object. Employed predominantly in disciplines like engineering and architecture, the CAD model enables designers to conceptualize, visualize, and analyze objects before their physical production. These models provide intricate visual representations of physical objects and facilitate the simulation of their operations across diverse scenarios. Moreover, there exists the possibility of amalgamating various methodologies, including equations, computer programs, 3D modeling, and CAD, to fashion more advanced digital twins capable of emulating complex systems with heightened fidelity.

In recent years, discussions surrounding digital twins have expanded into neuroscience and the broader healthcare sector, encompassing various projects related to pharmaceutical development and the digital replication of human body components. Through digital twins, it will soon become feasible to simulate disease effects on computers without initially intervening with the human body. This advancement opens avenues for *personalized* drug development tailored to specific patient conditions, aiming to thwart pathogenic processes while minimizing the side effects commonly associated with standard medications. Digital twins empower researchers to forecast a drug impact on the human brain and body, enable physicians to simulate surgeries or evaluate the outcomes of diagnostic procedures, and afford patients the foresight to anticipate how lifestyle changes may affect their health [55].

Digital twins of patients will not possess the characteristics of living beings, such as souls or emotions. They will serve as functional models of the brain-body-environment system. Digital twins could simulate how our health might change based on specific lifestyles or environmental factors. Perhaps the most intriguing aspect is their utility in swiftly evaluating new personalized therapies, free from ethical implications and offering substantial economic benefits. Consider, for instance, the scenario of determining the optimal dosage of a drug for a person with Parkinson's disease—the dosage that minimizes side effects while alleviating symptoms like tremors. Instead of subjecting the patient to discomfort, incurring high costs, and spending months on testing different drug dosages, we could expedite the process by testing various quantities on the digital twin of the Parkinsonian patient. The dosage that yields the most favorable outcome in the digital twin is likely to produce similar effects in the real patient. Consequently, clinicians will personalize therapy by tailoring it to the patient's unique characteristics, encapsulated in their digital twin (Fig. 4.1). Beyond medical conditions, the digital twin approach can be extended to basic research, facilitating the study of the functioning of various organisms.

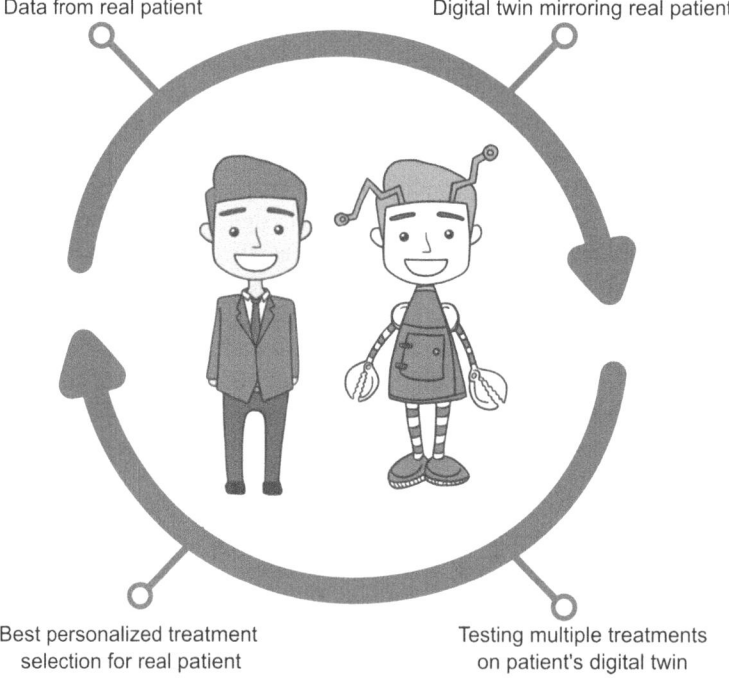

Fig. 4.1 Utilizing the digital twin to select the optimal therapy involves generating a model based on data from the real patient, including clinical, anatomo-physiological, and personal information. This model mirrors the patient's brain and body functions by creating a digital twin. Subsequently, clinicians could evaluate a range of therapies using the digital twin to pinpoint the most suitable personalized treatment for the specific patient. Figure adapted from [2]

This methodology promises to reduce and potentially eliminate the need for animal experimentation to evaluate the efficacy of new drugs. Instead, researchers could test drugs on their respective digital counterparts (cf. Sect. 4.3).

One method for developing digital twins involves employing a technique known as computational phenotyping [56]. Let's envision experimenting to investigate the motor behavior of subjects with Parkinson's disease. We might opt for a hand-eye coordination task, wherein participants are presented with a sequence of movements to replicate. Below are the steps to develop digital twins of patients using computational phenotyping.

1. Real-Subjects Data Collection. Gather data on the performance of patients participating in the experiment, focusing on factors such as fluidity and speed of movement execution concerning the assigned motor tasks. Additionally, data on subjects' neural activity during the experiment, potentially obtained through brain imaging techniques, can enhance the efficacy of computational phenotyping.
2. Digital Twin Equations. Construct a computational model capable of simulating interactions among the brain regions most pertinent to the task. For instance, if participants engage in a hand-eye coordination task, the computational model must elucidate, using mathematical equations, how specific brain areas involved in motor control and vision interact. This model comprises equations, such as differential equations, whose solutions depict how the activity of a particular brain region changes over time in response to other regions' activity. Some model parameters are predefined based on documented data. For instance, research literature might indicate that neurons in a brain region take 20 milliseconds to deactivate after stimulation, thus informing the model parameters representing neurons deactivation time. However, other parameters, termed "free parameters," require determination through statistical techniques. The model can be developed using software and subsequently evaluated through simulations conducted on a computer.
3. Model Free Parameters Identification. Employ statistical methods to approximate the values of the free parameters within the system of equations constituting the model, thereby aligning its behavior with that observed in subjects who participated in the experiment (data fitting). For example, statistical analyses might suggest that to reproduce the same performance in the hand-eye coordination task as observed in subjects, the free parameters p1 and p2 of the model should be set to 0.2 and 0.9, respectively. These parameters could symbolize, for instance, the concentration levels of two neuromodulators. Determining the values of the free parameters enables customization of the computational model to match the specific subject, transforming it into a digital twin capable of emulating the observed behavior. Given their biological significance, model parameters could allow us to make predictions regarding the neurophysiological processes underlying the observed behavior in real subjects.

4.3 Towards Eliminating Experiments Involving Animals

The digital twin can simulate the effects of pharmaceuticals and drug interactions within an organism. More broadly, digital twins allow the exploration of physiological processes underlying diseases through computer simulations, reducing the reliance on animal testing for evaluating new drugs. However, before therapies chosen via computational models can be applied clinically to humans, they must undergo additional experimental validations on both animals and humans. Consequently, it is not currently viable to entirely supplant the utilization of animals in medical research. Presently, computational models serve as a complementary tool to diminish the necessity for animal and human experimentation but not to entirely avoid these practices. Eliminating animals from research poses significant challenges. These obstacles include the complexity of the human organism, accounting for its interactions with the environment and life experiences, and the disparities between human beings and other animal species used in research. There are two computational ways to address these challenges and work toward a future without animal testing: the bottom-up and the top-down approaches.

The bottom-up approach necessitates computational models capable of creating digital twins with far greater detail than those currently feasible. This process requires deep expertise in biology, medicine, and physiology alongside vast datasets. The top-down approach aims to capture the essence of the organism-environment system using mathematical tools, prioritizing system-level understanding over detailed reproduction. The choice between these approaches is not straightforward, each presenting merits and drawbacks. The bottom-up method offers specificity and precision but demands substantial computational resources and may be challenging to analyze. The top-down approach simplifies complexities, enhancing computational efficiency and model comprehensibility. However, it needs careful consideration about what aspects to exclude from the model. A hybrid approach integrating bottom-up and top-down methods could offer a viable path forward. By combining detailed models of crucial components with system-level models, researchers can exploit the benefits of each technique. However, regardless of the chosen approach, rigorous standards for validating simulations with computational models as digital twins are imperative to ensure accuracy and reproducibility. Fostering collaboration among scientists from diverse disciplines, software engineers, and ethics and regulatory experts is essential. Additionally, synergies between computational models and emerging technologies, such as organ-on-a-chip devices hold promise for reducing reliance on animal testing. Consequently, the ongoing advancements in these fields hint at an ethical and effective alternative to animal usage in scientific research.

4.4 The Four-Step Method

Understanding the intricate dynamics of the brain remains a complex, cross-disciplinary challenge that extends beyond neuroscience. A promising approach tackles this challenge by combining computational methods with biological insights. This section presents a four-step methodology to create system-level brain models using ODEs and Python. The first step in our process is to create a block diagram-a foundational map of the brain regions to study. This stage involves developing a schematic representation of the interconnected brain areas that form the network under study. Mapping out this circuitry helps clarify the system core structure and lays the groundwork for the subsequent mathematical framework. The second step begins with scripting ODEs, which serve as the mathematical framework for modeling the dynamic evolution of events in complex systems. In brain modeling, ODEs take on the role of quantifying neuronal and synaptic dynamics. ODEs could integrate variables like neuron membrane potential, synaptic conductance, and other critical factors. This integration forms the foundation for capturing the intricate interplay of these variables, creating a dynamic portrait of brain circuit behavior. The third step involves translating these equations into Python. Python serves as the ideal platform for implementing ODE-based brain models. In particular, this phase focuses on converting mathematical equations into code, adjusting model parameters, and running simulations to uncover insights into the intricate dynamics governing brain circuits function. In the final step, the focus shifts to refining model parameters through computational phenotyping techniques. Achieving verisimilitude in replicating real-world brain dynamics requires fine-tuning parameters until they align closely with empirical data. Here, optimization techniques-whether through gradient-based methods or genetic algorithms-play a pivotal role, iteratively adjusting the model core to minimize the gap between simulated results and observed data.

This four-step method gives a comprehensive guide for developing system-level brain models through an agile fusion of ODEs and Python tools. Each step takes researchers further from abstract theory to concrete data, unraveling the complexities of brain circuit dynamics. This approach could enable researchers to deeply explore the brain circuits complex behaviors and bridge the challenging gap between theoretical insights and empirical observations. The example below outlines the four-step process for implementing a system-level model to investigate the cerebellum role in Alzheimer's disease. Recent findings have brought attention to the cerebellum's involvement in Alzheimer's pathology despite its traditional exclusion from the core understanding of the disease. In the next section, further examples will showcase specific aspects and applications of the four-step method, focusing on brain learning mechanisms, brain-body interactions, model stability, and simulations of potential therapeutic interventions. Except for the last example, drawn from a scientific article authored by the researchers, the other examples will simplify the discussion by omitting units of measurement for the parameters used in the models. This approach helps avoid complicating the analysis with unit conversions. For more information on this topic, please refer to the final example.

4.4.1   Step 1: Drawing the System-Level Model

The first step aims to create a schematic representation of the interconnected brain regions relevant to studying the cerebellum role in Alzheimer's disease. This neurodegenerative disorder significantly affects the brain, leading to a cascade of challenges involving memory, cognition, and behavior. As the most common neurodegenerative condition, Alzheimer's is characterized by the gradual degeneration of nerve cells in the brain, with the prefrontal cortex playing a crucial role in this process [57]. While there is no cure, treatments are available to manage symptoms and enhance the quality of life for those affected. Traditionally, researchers associate the cerebellum with motor functions, but its intricate connections reveal its involvement in cognitive processes [58, 59]. Recent studies highlight a possible link between the cerebellum and Alzheimer's, opening new avenues for innovative and forward-thinking scientific research [60].

An essential starting point is to conduct an in-depth literature review to develop a strong circuit hypothesis and construct a clear block diagram representing the potential neural system underlying the cerebellum involvement in Alzheimer's disease. A good approach includes examining critical review articles to gain a comprehensive overview. In this process, authoritative sources, such as a research article published by a leading group of cerebellum experts, can provide valuable insights. It is also worth considering studies that, while not directly linking the cerebellum to the brain regions involved in Alzheimer's disease, still provide evidence for its potential indirect involvement. A system-level approach proves valuable here, as exploring research across different pathologies can yield insights. In this regard, one fascinating study examines the cerebellum role in autism. This research is relevant to our work as it highlights a neural circuit connecting the cerebellum to the prefrontal cortex through the ventral tegmental area (VTA), suggesting that the cerebellum may influence prefrontal cortex activity. This connection is significant because the prefrontal cortex, with its function in working memory, is severely impacted by Alzheimer's disease. Working memory temporarily stores and manipulates information, enabling complex tasks such as decision-making, problem-solving, and planning. The accumulation of amyloid plaques and tau tangles, hallmarks of the disease, disrupts the prefrontal cortex early in the disease progression, negatively affecting working memory. Consequently, individuals with Alzheimer's often struggle with organizing thoughts, making decisions, and performing routine tasks.

Based on this perspective, we propose a block diagram illustrating a potential connection between the cerebellum and the prefrontal cortex via the VTA, which plays a critical role in dopamine release. This schema lays the groundwork for constructing an initial neural architecture to explore the cerebellum role in Alzheimer's disease (the first circuit hypothesis). It is a preliminary framework that provides a simplified model, with opportunities for refinement-such as incorporating additional regions or enhancing the equations governing brain regions interactions-detailed in the subsequent step 2. The level of refinement will depend on the research objectives and the desired complexity of the model. As the model evolves through ongoing

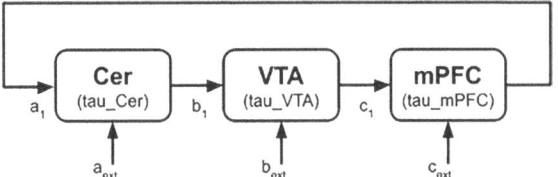

Fig. 4.2 Block diagram depicting a potential interaction between the cerebellum (Cer), ventral tegmental area (VTA), and medial prefrontal cortex (mPFC). This interaction is governed by parameters a_1 and b_1, with τ representing the neural decay times for each region. Additionally, parameters a_{ext}, b_{ext}, and c_{ext} capture the influence of the broader brain context on these three areas

literature review and further development, this initial circuit may expand to include additional brain regions and more sophisticated mechanisms.

The simplified diagram in Fig. 4.2 assumes that the cerebellum (Cer) connects to the ventral tegmental area (VTA), which, in turn, projects to the medial prefrontal cortex (mPFC), with the mPFC completing the loop by connecting back to the cerebellum. All connections are excitatory, with their strengths represented by the numerical parameters a_1, b_1, and c_1. The parameters a_{ext}, b_{ext}, and c_{ext} reflect the influence of the broader brain context on these three regions. In addition, the block schema includes three relaxation terms (τ) with a negative sign representing the decay time—the period neurons take to deactivate after stimulation. Rather than attempting to simulate the entire brain, the model simplifies and integrates relevant aspects, considering the potential impact of external brain areas. Although these areas may be less directly involved in Alzheimer's disease, they could still modulate the functions of the cerebellum, VTA, and mPFC.

4.4.2 Step 2: Writing the Ordinary Differential Equations

Starting with the diagram in Fig. 4.2, derive the system of differential equations representing the model. This involves writing down the balance equations (see Sect. 3.2) for each block depicted in the diagram.

$$\dot{Cer} = -\tau_{Cer} \cdot Cer + a_1 \cdot mPFC + a_{ext} \tag{4.4.1}$$

$$\dot{VTA} = -\tau_{VTA} \cdot VTA + b_1 \cdot Cer + b_{ext} \tag{4.4.2}$$

$$\dot{mPFC} = -\tau_{mPFC} \cdot mPFC + c_1 \cdot VTA + c_{ext} \tag{4.4.3}$$

These equations propose a potential mechanism by which the cerebellum influences the medial prefrontal cortex through an indirect pathway involving the VTA and dopamine. This forms the core hypothesis of the model.

There are nine parameters in total. How can their values be determined? Existing literature could provide values for some parameters, such as decay constants. For other parameters, like those that summarize the influence of one brain region on another, the literature may not offer direct values. In such cases, optimization techniques, such as genetic algorithms, can assist in estimating these free parameters. The literature could also give reference values for average activity levels in the cerebellum, VTA, and mPFC under specific resting conditions in healthy subjects and patients. These values could serve as target activities and set the initial values. Optimization techniques adjust the free parameters to match these target activity levels in the model. When the number of free parameters is manageable, manual adjustment through trial and error can also achieve the target values for the brain regions. The following sections will detail these processes further.

4.4.3   Step 3: Building the Model with Python

At this point, the next step involves implementing the model described by the differential equations in Python. Python provides a convenient method for solving these equations, allowing the determination of how neural activities (dependent variables) evolve over time based on the values of model parameters (independent variables). The following sections will outline the various components of the Python code, with each block serving a specific purpose.

The first code block imports the necessary Python libraries. The solve_ivp function from the scipy.integrate module handles the solution of the system of differential equations representing the model. This function addresses ODEs as initial value problems (cf., Sect. 2.8), meaning it calculates how the system evolves over time from specified initial conditions. In essence, solve_ivp processes the system of differential equations and initial conditions to return a numerical solution across a defined time interval. It employs various numerical integration techniques to approximate the solution.

```
1   # Import required libraries for numerical operations,
    ↪   plotting, and solving differential equations
2   import numpy as np                    # For numerical
    ↪   computations and array operations
3   import matplotlib.pyplot as plt       # For creating plots
    ↪   and visualizations
4   import matplotlib                     # For additional
    ↪   plotting functionalities
5   from scipy.integrate import solve_ivp  # For solving
    ↪   systems of ordinary differential equations (ODEs)
```

The next step involves developing a Python function to encapsulate the ODEs that represent the computational model. This function will work as input for the solve_ivp function from the SciPy library, which numerically solves the ODE system. To implement this, define the model function to accept at least two arguments: t (the independent variable representing time) and x (the dependent variable representing the state vector, i.e., the activity of Cer, VTA and mPFC). The function then returns the derivatives of x with respect to t. Along with t and x, the model function also requires the model parameters as additional arguments. The function returns an array-like object containing the derivatives of x with respect to t.

```
1   # Define the model function to represent the system of
    ↪  differential equations
2   def model(t, x, a_ext, a_1, tau_Cer, b_ext, b_1,
    ↪  tau_VTA, c_ext, c_1, tau_mPFC):
3       Cer  = x[0]  # Firing rate of cerebellar neurons (Hz)
4       VTA  = x[1]  # Firing rate of ventral tegmental area
        ↪  neurons (Hz)
5       mPFC = x[2]  # Firing rate of medial prefrontal
        ↪  cortex neurons (Hz)
6
7       # Compute the rate of change for each variable
8       dCerdt  = a_ext + a_1 * mPFC - tau_Cer * Cer
9       dVTAdt  = b_ext + b_1 * Cer - tau_VTA * VTA
10      dmPFCdt = c_ext + c_1 * VTA - tau_mPFC * mPFC
11
12      return [dCerdt, dVTAdt, dmPFCdt]
```

In the function:

- t represents the current time (scalar), which serves as the independent variable in the ODE system. The solve_ivp function integrates the system over a specified range of time values.
- x is an array indicating the current state of the system. For this model, x is a one-dimensional array with three elements: Cer, VTA, and mPFC. These represent the firing rates of cerebellar neurons, ventral tegmental area neurons, and medial prefrontal cortex neurons, respectively.
- a_ext, a_1, a_2, b_ext, b_1, b_2, c_ext, c_1, c_2 are the parameters of the model. These constants define the system dynamics and influence how the firing rates evolve over time based on the interactions described by the ODEs.

The model function calculates the derivatives of the state variables (dCerdt, dVTAdt, dmPFCdt) at a given time t based on the current state x and the model parameters.

When using solve_ivp, the model function must be supplied as one of its arguments. Additionally, provide the initial conditions (x0), the time span for integration (t_span), and any extra parameters required by the model function. For this model, these additional parameters are a_ext, a_1, a_2, b_ext, b_1, b_2, c_ext, c_1, and c_2. Subsequent Python code block will define x0 and these parameters.

```
1  # Define the initial conditions for the model state
   ↪  variables
2  Cer_init = 4.0  # Initial firing rate of cerebellar
   ↪  neurons (Hz)
3  VTA_init = 3.0  # Initial firing rate of ventral
   ↪  tegmental area neurons (Hz)
4  mPFC_init = 5.0 # Initial firing rate of medial
   ↪  prefrontal cortex neurons (Hz)
5  x0 = [Cer_init, VTA_init, mPFC_init]  # Initial state
   ↪  vector for the system
```

The initial conditions represent the average activity levels in healthy subjects across the three brain regions considered in the model. These values are sometimes available in the scientific literature. However, when precise data is difficult to obtain, reasonable estimates can be made based on typical neural activity ranges for different brain areas. This study follows such an approach. While explicit average firing rates (in Hz) for the cerebellum, VTA, and mPFC are not always directly mentioned in existing literature, general insights can still be drawn. For instance, it is well-established that synchronization of neuronal oscillations below approximately 100 Hz is a common feature of brain activity, playing a crucial role in communication within neural circuits [61, 62]. Cerebellar neural activity varies depending on the task. Typically, the cerebellum displays oscillatory activity within the 4–30 Hz range. In healthy individuals at rest, neural activity tends to fall between 8 and 12 Hz, reflecting a relaxed but alert state. However, during motor or cognitive tasks, cerebellar activity can increase considerably, often exceeding 20 Hz [58, 63]. Neural activity in the mPFC often exhibits oscillatory patterns within the theta (4–8 Hz) and gamma (30–100 Hz) frequency ranges. Theta oscillations are particularly associated with memory encoding and retrieval, while fast gamma oscillations, above 30 Hz, play a key role in active cortical processing within the mPFC [64, 65]. The VTA is a key part of the brain reward system, central to motivation, reinforcement learning, and addiction. VTA neurons display spontaneous activity and phasic firing patterns. The firing rates of dopaminergic neurons in the VTA vary significantly, with a baseline tonic rate typically between 1-10 Hz, while bursts of activity can surpass 20 Hz [66].

In addition to the initial values, consider target values representing the activities of a specific healthy individual. The first objective in this initial phase is to develop a model that can reproduce these target values.

```
1   # Set the target firing rates for each brain region for
    ↪    a specific subject
2   Cer_target = 6.0   # Firing rate of cerebellar neurons
    ↪    (Hz)
3   VTA_target = 5.0   # Firing rate of ventral tegmental
    ↪    area neurons (Hz)
4   mPFC_target = 7.0 # Firing rate of medial prefrontal
    ↪    cortex neurons (Hz)
5   xT = [Cer_target, VTA_target, mPFC_target]   # Target
    ↪    state vector for the system
```

Establishing initial parameter values for the model is just as important as defining the initial conditions. While certain parameters, like neuron decay constants in specific brain regions, can be reliably obtained from existing research, many others—particularly those representing inter-area influences—are often assigned random values within a biologically plausible range due to a lack of available data. This randomness reflects the complexity of identifying precise values. Inter-area influences, for example, can be affected by various factors, making it difficult to derive a single value from the literature. Neurotransmitter concentrations, for instance, can modulate the strength and efficacy of connections between brain regions. Additionally, neuron firing patterns and synchronization within a particular area can impact communication with other regions. The density and strength of white matter tracts connecting different areas also significantly shape information flow and regulate interactions between brain regions. For simplicity, random values will be used for all the parameters below.

```
1    # Setting the initial values for the parameters
2
3    # Cerebellum (Cer)
4    a_ext = 1.4   # External input to the cerebellum
5    a_1   = 0.2   # Influence of the medial prefrontal cortex
     ↪    (mPFC) on the cerebellum
6    a_2   = 0.8   # Neuronal decay constant in the cerebellum
7
8    # Ventral Tegmental Area (VTA)
9    b_ext = 0.85 # External input to the ventral tegmental
     ↪    area
10   b_1   = 0.25 # Influence of the cerebellum on the VTA
11   b_2   = 1.0  # Neuronal decay constant in the VTA
12
13   # Medial Prefrontal Cortex (mPFC)
14   c_ext = 0.4   # External input to the mPFC
15   c_1   = 0.8   # Influence of the VTA on the mPFC
16   c_2   = 0.2   # Neuronal decay constant in the mPFC
```

Before solving the system of ordinary differential equations using the solve_ivp function, it is necessary to define the t_span parameter. It determines the granularity

of the numerical integration, specifying the level of detail in the simulation. The
solve_ivp solver will compute the solution for t within the t_span, a tuple (t0, tf) that
defines the initial and final times for integration.

```
1   # Define the start (t0) and end (tf) times for the
    ↪ simulation
2   t_span = (0, 30)
```

The ODEs system can now be solved using the solve_ivp function. The first argument is the model function, which defines the ODE system. The second argument,
t_span, sets the time interval for the solution, and the third argument, x0, specifies
the initial values of the dependent variables, representing the system initial conditions at t0. The args parameter passes additional arguments to the model function
as a tuple containing any extra parameters required by the ODE model. This argument becomes necessary when modifying model parameters. The solve_ivp function
allows to customize the simulation by selecting different options. For example, the
method argument determines the integration method to use, with methods offering
various trade-offs in accuracy, stability, and efficiency. For more detailed information,
refer to the solve_ivp documentation digiting help(solve_ivp).

```
1   # Solving the ODEs system
2   solution = solve_ivp(
3       model,                  # The callable model function
        ↪ defining the ODE system
4       t_span,                 # Time span for integration
        ↪ (start time and end time)
5       x0,                     # Initial values of the dependent
        ↪ variables
6       args=(a_ext, a_1, a_2, b_ext, b_1, b_2, c_ext, c_1,
        ↪ c_2)  # Additional parameters for the model
7   )
```

Extract the contents of the solution array and assign them to separate objects for
easier manipulation and graphical representation.

```
1   # Extract and assign solution components
2   t    = solution.t           # Time points at which the
    ↪ solution was evaluated
3   Cer  = solution.y[0]        # Output for the cerebellum
4   VTA  = solution.y[1]        # Output for the ventral
    ↪ tegmental area
5   mPFC = solution.y[2]        # Output for the medial
    ↪ prefrontal cortex
```

The final step is to plot the results.

```python
# Define font sizes for plotting
SMALL_SIZE  = 20
MEDIUM_SIZE = 22

# Set default text sizes
plt.rc('font', size=SMALL_SIZE)             # Font size for
  ↪  default text
plt.rc('axes', titlesize=SMALL_SIZE)        # Font size for
  ↪  axis titles
plt.rc('axes', labelsize=MEDIUM_SIZE)       # Font size for
  ↪  axis labels
plt.rc('xtick', labelsize=SMALL_SIZE)       # Font size for
  ↪  x-axis tick labels
plt.rc('ytick', labelsize=SMALL_SIZE)       # Font size for
  ↪  y-axis tick labels
plt.rc('legend', fontsize=MEDIUM_SIZE)      # Font size for
  ↪  legend
```

```python
plt.figure(1)

# Plot the results for each component with distinct
  ↪  colors and line widths
plt.plot(t, Cer,  'b', linewidth=2, label='Cer')
plt.plot(t, VTA,  'g', linewidth=2, label='VTA')
plt.plot(t, mPFC, 'r', linewidth=2, label='mPFC')

# Add horizontal lines for target values, limited to the
  ↪  time range
plt.plot([0, t[-1]], [Cer_target, Cer_target], 'b--',
  ↪  linewidth=1, label='Target Cer')
plt.plot([0, t[-1]], [VTA_target, VTA_target], 'g--',
  ↪  linewidth=1, label='Target VTA')
plt.plot([0, t[-1]], [mPFC_target, mPFC_target], 'r--',
  ↪  linewidth=1, label='Target mPFC')

# Label the axes
plt.xlabel('Time')
plt.ylabel('Values')

# Add a legend outside the plot
plt.legend(loc='upper left', bbox_to_anchor=(1, 1))
plt.grid(True)

# Display the plot
plt.show()
```

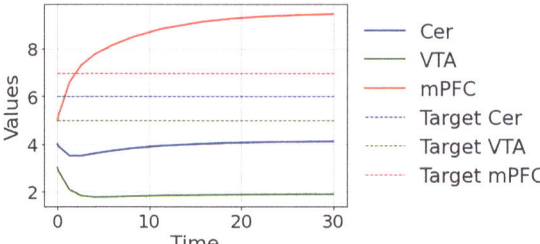

The plot illustrates the model output for Cer, VTA, and mPFC neurons, compared with the target values for a healthy subject. The model, using random parameters, fails to match these target values, indicating that the current parameters do not accurately reflect the expected physiological activity of the healthy subject. The next step will show the process to obtain the parameters that reproduce the specific subject data.

4.4.4   Step 4: Optimizing Model Parameters to Match Subject-Specific Data

This section explores how to use genetic algorithms (GAs) to determine the parameters that enable the model to accurately reproduce target data for healthy subjects. GAs provide a powerful method for finding optimal parameters for models by simulating the process of natural selection (see Sect. 2.11.3). The approach begins with the **initialization** phase. This involves generating a diverse set of potential solutions, referred to as chromosomes. Each chromosome represents a unique set of parameters for the model being optimized. Next, in the **evaluation** phase, each set of parameters is assessed using a fitness function. This function measures how effectively the model performs with the given parameters by comparing its output to established target values. This assessment helps to determine which solutions are most promising. Following evaluation, the **selection** phase comes into play. Here, the top-performing chromosomes-those with the highest fitness scores-are chosen to contribute to the next generation. This ensures that the most effective solutions are carried forward and refined. The **crossover** phase involves creating new solutions by combining parts of the selected chromosomes. This process mimics genetic recombination seen in nature and introduces new combinations of parameters. The goal is to generate potentially better solutions by merging successful traits from different chromosomes. To further enhance the search, the **mutation** phase introduces small, random changes to some of the new solutions. This step maintains genetic diversity within the population, which helps to prevent the algorithm from becoming stuck in local optima. The mutation rate determines how frequently these random changes occur. In the **replacement** phase, the old population of solutions is replaced with the newly generated ones. This step ensures that each generation evolves and improves based on the previous selection and mutation processes. Finally, the **iteration** phase repeats these steps for a specified number of generations or until the algorithm converges on an optimal set of parameters. The total number of generations defines the duration of the algorithm run and the extent to which it explores the solution space.

The Python examples below show the use of mutation alone in the GA. This approach is more similar to Differential Evolution (DE) techniques (see Sect. 2.11.3.1) that relies on differential mutation and recombination rather than crossover and mutation as in GA. This often allows DE to converge more quickly on optimal solutions due to its direct method of combining differences between individuals. DE typically requires fewer parameters to be tuned compared to GA, which might simplify its application in some scenarios. Both genetic algorithms and differential evolution are effective for parameter optimization. The choice between them depends on the specific problem, the need for exploration versus exploitation, and available computational resources.

Below are the critical parameters used in the GA to optimize our model parameters.

- **Number of Chromosomes in the Population (Population Size)**: This parameter dictates the number of potential solutions evaluated at each generation. A larger population explores a wider range of solutions but requires more computational resources.
- **Mutation Rate**: This controls the frequency of random changes introduced into the solutions. A higher mutation rate explores new areas of the solution space but may disrupt the search process.
- **Number of Survivors per Generation (Elitism)**: This ensures that the best-performing solutions are carried over to the next generation, preserving high-quality solutions and preventing them from being lost.
- **Total Number of Generations**: This defines the number of iterations the algorithm will run. More generations allow for a more thorough search but require additional computation time.

For clarity, the complete code, including the initial section discussed in Step 3, will be provided below. The code sets the random seed for both the random module and NumPy to ensure that random number generation is reproducible. Using the same seed value (1 in this case) guarantees that the results of random operations will be consistent across multiple runs of the program.

```
1   # Import required libraries for numerical operations, plotting, and
    ↪   solving differential equations
2   import random                         # For generating random numbers
    ↪   and random operations
3   import numpy as np                     # For numerical computations and
    ↪   array operations
4   import matplotlib.pyplot as plt        # For creating plots and
    ↪   visualizations
```

```python
5   import matplotlib                         # For additional plotting
    ↪   functionalities
6   from scipy.integrate import solve_ivp  # For solving systems of ODEs
7   import copy  # Import the copy module for creating deep copies of
    ↪   objects
8
9   # Set seed for reproducibility
10  random.seed(1)
11  np.random.seed(1)
12
13  # Define the system of ordinary differential equations (ODEs) that
    ↪   model neural firing rates
14  def model(t, x, a_ext, a_1, tau_Cer, b_ext, b_1, tau_VTA, c_ext,
    ↪   c_1, tau_mPFC):
15      # Unpack the state variables representing firing rates of
        ↪   different brain regions
16      Cer  = x[0]  # Firing rate of cerebellar neurons (Hz)
17      VTA  = x[1]  # Firing rate of ventral tegmental area neurons
        ↪   (Hz)
18      mPFC = x[2]  # Firing rate of medial prefrontal cortex neurons
        ↪   (Hz)
19
20      # Define the ODEs governing the interactions between the regions
21      dCerdt  = a_1 * mPFC - tau_Cer * Cer + a_ext   # Change in
        ↪   cerebellar firing rate
22      dVTAdt  = b_1 * Cer - tau_VTA * VTA + b_ext    # Change in VTA
        ↪   firing rate
23      dmPFCdt = c_1 * VTA - tau_mPFC * mPFC + c_ext  # Change in
        ↪   mPFC firing rate
24
25      # Return the derivatives as a list to be used by the solver
26      return [dCerdt, dVTAdt, dmPFCdt]
27
28  # Initial conditions for the system (firing rates of Cer, VTA, and
    ↪   mPFC at t=0)
29  Cer_init  = 4.0   # Initial cerebellar firing rate (Hz)
30  VTA_init  = 3.0  # Initial VTA firing rate (Hz)
31  mPFC_init = 5.0  # Initial mPFC firing rate (Hz)
32  x0 = [Cer_init, VTA_init, mPFC_init]  # Combine into initial state
    ↪   vector
33
34  # Model parameters for each equation
35  a_ext, a_1, tau_Cer = 1.4, 0.2, 0.8  # Parameters influencing
    ↪   cerebellar neurons
36  b_ext, b_1, tau_VTA = 0.85, 0.25, 1.0  # Parameters influencing VTA
    ↪   neurons
37  c_ext, c_1, tau_mPFC = 0.4, 0.2, 0.8  # Parameters influencing mPFC
    ↪   neurons
38
39  # Time parameters for the simulation
40  n_step = 10001  # Number of time steps to evaluate
41  t_span = [0, 30]  # Time span for the simulation (0 to 50 units of
    ↪   time)
42  t_eval = np.linspace(t_span[0], t_span[1], n_step)  # Time points
    ↪   at which to evaluate the solution
43
44  # Target steady-state values for the healthy system (used in the
    ↪   genetic algorithm)
```

```
45   Cer_target = 6.0    # Firing rate of cerebellar neurons (Hz)
46   VTA_target = 5.0   # Firing rate of ventral tegmental area neurons
     ↪    (Hz)
47   mPFC_target = 7.0 # Firing rate of medial prefrontal cortex neurons
     ↪    (Hz)
48   target_values = [Cer_target, VTA_target, mPFC_target] # Desired
     ↪    steady-state firing rates for Cer, VTA, and mPFC
49
50   # Genetic Algorithm (GA) parameters for optimizing the model
     ↪    parameters
51   chromo_pop_num = 100   # Number of chromosomes in the population
     ↪    (population size)
52   mut_rate = 0.01        # Mutation rate for introducing small random
     ↪    changes to genes
53   surv_num = 1           # Number of survivors per generation (elitism)
54   gen_num = 70           # Total number of generations for the GA
55
56   # Initialize the first generation of genes (model parameters to be
     ↪    optimized)
57   first_gene_values_list = [a_ext, a_1, tau_Cer, b_ext, b_1, tau_VTA,
     ↪    c_ext, c_1, tau_mPFC]
58
59   # Solve the ODE system with the initial parameters to evaluate
     ↪    fitness before GA starts
60   sol = solve_ivp(model, t_span, x0,
     ↪    args=tuple(first_gene_values_list), t_eval=t_eval)
61
62   # Calculate the fitness based on the steady-state error (difference
     ↪    from target values)
63   predicted = np.array([sol.y[0, -1], sol.y[1, -1], sol.y[2, -1]])  #
     ↪    Predicted final values
64   errors = predicted - np.array(target_values)  # Difference from
     ↪    target
65   abs_perc_errors = np.abs((errors / predicted) * 100)  # Absolute
     ↪    percentage error
66   mean_ape = np.mean(abs_perc_errors)  # Mean absolute percentage
     ↪    error (MAPE)
67   start_fitness = (100 - mean_ape) / 100  # Fitness value (higher is
     ↪    better)
68
69   # Initialize the best fitness value and the selected population for
     ↪    the first generation
70   best_total_fitness = start_fitness  # Track the best fitness
     ↪    achieved so far
71   selected_pop = first_gene_values_list.copy()  # Set the initial
     ↪    best parameters as the selected population
72
73   # Store the fitness history to track progress across generations
74   total_fitness_history = []
75
76   # Genetic Algorithm main loop to evolve the population over
     ↪    multiple generations
77   for gen in range(gen_num):
78
79       # Create a new population by applying random mutations to the
     ↪        best solution from the previous generation
80       chromo_pop = []
81       for _ in range(chromo_pop_num):
```

```
82              # Apply a small mutation to each gene in the selected
           ↪    population
83              new_chromo = [gene + np.random.normal(0.0, gene * mut_rate)
           ↪    for gene in selected_pop]
84              chromo_pop.append(new_chromo)
85
86          # Evaluate the fitness of each chromosome in the population
87          pop_total_fitness = []
88          for chromo in chromo_pop:
89              # Solve the ODE system with the chromosome's parameters to
           ↪    calculate fitness
90              sol = solve_ivp(model, t_span, x0, args=tuple(chromo),
           ↪    t_eval=t_eval)
91
92              # Calculate the steady-state values for this chromosome
93              predicted = np.array([sol.y[0, -1], sol.y[1, -1], sol.y[2,
           ↪    -1]])
94
95              # Compute fitness based on how closely the solution matches
           ↪    the target values (MAPE)
96              errors = predicted - np.array(target_values)
97              abs_perc_errors = np.abs((errors / predicted) * 100)
98              mean_ape = np.mean(abs_perc_errors)
99              fitness = (100 - mean_ape) / 100  # Fitness value (higher
           ↪    is better)
100
101             pop_total_fitness.append(fitness)  # Append fitness of this
           ↪    chromosome to the population's fitness list
102
103         # Sort the population by fitness (highest first) and select the
           ↪    best chromosome
104         sorted_fitness_pop = sorted(zip(pop_total_fitness, chromo_pop),
           ↪    reverse=True)
105
106         # Update the best fitness and selected population for the next
           ↪    generation
107         best_total_fitness = sorted_fitness_pop[0][0]  # Best fitness
           ↪    in this generation
108         selected_pop = sorted_fitness_pop[0][1]  # Corresponding
           ↪    chromosome with the best fitness
109
110         # Store the best fitness value in the fitness history for
           ↪    future analysis
111         total_fitness_history.append(best_total_fitness)
112
113     # Plot the evolution of fitness over the generations
114     plt.figure(figsize=(10, 6))
115     plt.plot(total_fitness_history, label="Best Fitness Over
           ↪    Generations", marker='o')
116     plt.xlabel('Generation')  # Label for x-axis
117     plt.ylabel('Best Fitness')  # Label for y-axis
118     plt.title('Evolution of Fitness Over Generations')  # Title of the
           ↪    plot
119     plt.legend()  # Show legend on the plot
120     plt.grid(True)  # Enable grid for better readability
121     plt.show()
122
123     # Solve the ODE system using the best parameters found by the
           ↪    genetic algorithm
```

```python
124    sol = solve_ivp(model, t_span, x0, args=tuple(selected_pop),
       ↪  t_eval=t_eval)
125
126    # Plot the results of the simulation with the evolved parameters
127    plt.figure(1)
128
129    plt.plot(sol.t, sol.y[0], 'b', linewidth=2, label='Cer')
130    plt.plot(sol.t, sol.y[1], 'g', linewidth=2, label='VTA')
131    plt.plot(sol.t, sol.y[2], 'r', linewidth=2, label='mPFC')
132
133    # Add horizontal lines for target values, limited to the time range
134    plt.plot([0, sol.t[-1]], [Cer_target, Cer_target], 'b--',
       ↪  linewidth=1, label='Target Cer')
135    plt.plot([0, sol.t[-1]], [VTA_target, VTA_target], 'g--',
       ↪  linewidth=1, label='Target VTA')
136    plt.plot([0, sol.t[-1]], [mPFC_target, mPFC_target], 'r--',
       ↪  linewidth=1, label='Target mPFC')
137
138    # Label the axes
139    plt.xlabel('Time')
140    plt.ylabel('Values')
141
142    # Add a legend outside the plot
143    plt.legend(loc='upper left', bbox_to_anchor=(1, 1))
144    plt.grid(True)
145
146    # Display the plot
147    plt.show()
148
149    # Display the parameters of the healthy subject identified by the
       ↪  genetic algorithm
150    formatted_params = [f"{param:.2f}" for param in selected_pop]
151
152    # Print parameters in specified order
153    print("Parameters of the healthy subject identified by the genetic
       ↪  algorithm:")
154    print(f"a_ext: {formatted_params[0]}")
155    print(f"a_1: {formatted_params[1]}")
156    print(f"tau_Cer: {formatted_params[2]}")
157    print(f"b_ext: {formatted_params[3]}")
158    print(f"b_1: {formatted_params[4]}")
159    print(f"tau_VTA: {formatted_params[5]}")
160    print(f"c_ext: {formatted_params[6]}")
161    print(f"c_1: {formatted_params[7]}")
162    print(f"tau_mPFC: {formatted_params[8]}")
```

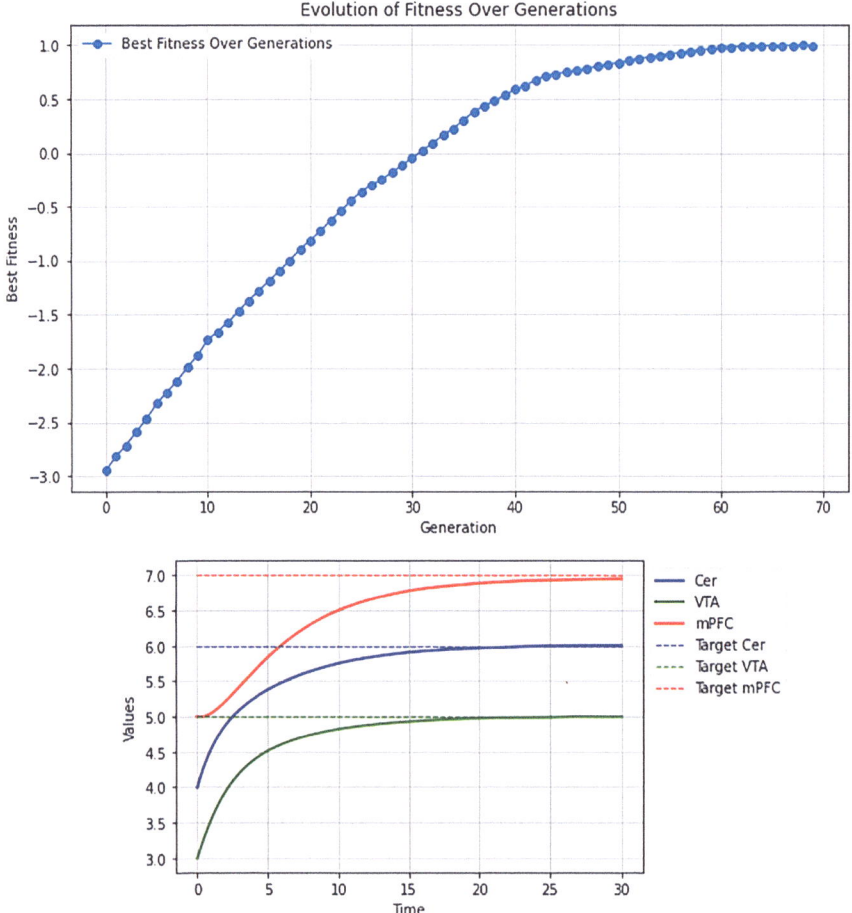

```
1   Parameters of the healthy subject identified by the
    ↪  genetic algorithm:
2   a_ext: 1.88
3   a_1: 0.21
4   tau_Cer: 0.56
5   b_ext: 1.27
6   b_1: 0.32
7   tau_VTA: 0.64
8   c_ext: 0.53
9   c_1: 0.33
10  tau_mPFC: 0.31
```

The algorithm achieves a fitness score of approximately 100%. This result demonstrates that the genetic algorithm effectively identifies parameters that enable the model to approximate the target data with high accuracy. These parameters concisely represent the healthy subject model.

4.4.5 Transition from Healthy Subject to Alzheimer's Disease

Starting with the healthy subject model (i.e., using the parameters representing the healthy system as initial values), it is possible to simulate data for a subject with Alzheimer's disease. For instance, assuming that data indicate Alzheimer's disease causes mPFC underactivation to as low as 4 Hz, the genetic algorithm can be run to identify new model parameters that align with this target mPFC data.

```python
1   # Import required libraries for numerical operations,
    ↪  plotting, and solving differential equations
2   import random                          # For generating
    ↪  random numbers and random operations
3   import numpy as np                      # For numerical
    ↪  computations and array operations
4   import matplotlib.pyplot as plt     # For creating plots
    ↪  and visualizations
5   import matplotlib                       # For additional
    ↪  plotting functionalities
6   from scipy.integrate import solve_ivp  # For solving
    ↪  systems of ODEs
7   import copy  # Import the copy module for creating deep
    ↪  copies of objects
8
9   # Set seed for reproducibility
10  random.seed(1)
11  np.random.seed(1)
12
13  # Define the system of ordinary differential equations
    ↪  (ODEs) that model neural firing rates
14  def model(t, x, a_ext, a_1, tau_Cer, b_ext, b_1,
    ↪  tau_VTA, c_ext, c_1, tau_mPFC):
15      # Unpack the state variables representing firing
        ↪  rates of different brain regions
16      Cer  = x[0]  # Firing rate of cerebellar neurons
        ↪  (Hz)
17      VTA  = x[1]  # Firing rate of ventral tegmental
        ↪  area neurons (Hz)
18      mPFC = x[2]  # Firing rate of medial prefrontal
        ↪  cortex neurons (Hz)
19
20      # Define the ODEs governing the interactions
        ↪  between the regions
21      dCerdt  = a_1 * mPFC - tau_Cer * Cer + a_ext   #
        ↪  Change in cerebellar firing rate
22      dVTAdt  = b_1 * Cer - tau_VTA * VTA + b_ext    #
        ↪  Change in VTA firing rate
```

```
23      dmPFCdt = c_1 * VTA - tau_mPFC * mPFC + c_ext    #
        ↪    Change in mPFC firing rate
24
25         # Return the derivatives as a list to be used by
        ↪    the solver
26      return [dCerdt, dVTAdt, dmPFCdt]
27
28   # Initial conditions for the system (firing rates of
     ↪    Cer, VTA, and mPFC at t=0)
29   Cer_init  = 6.0    # Initial cerebellar firing rate (Hz)
30   VTA_init  = 5.0   # Initial VTA firing rate (Hz)
31   mPFC_init = 7.0   # Initial mPFC firing rate (Hz)
32   x0 = [Cer_init, VTA_init, mPFC_init]   # Combine into
     ↪    initial state vector
33
34   # Healthy model parameters for each equation
35   a_ext, a_1, tau_Cer = 1.76, 0.21, 0.54   # Parameters
     ↪    influencing cerebellar neurons
36   b_ext, b_1, tau_VTA = 1.12, 0.29, 0.58   # Parameters
     ↪    influencing VTA neurons
37   c_ext, c_1, tau_mPFC = 0.59, 0.32, 0.31   # Parameters
     ↪    influencing mPFC neurons
38
39   # Time parameters for the simulation
40   n_step = 10001   # Number of time steps to evaluate
41   t_span = [0, 30]   # Time span for the simulation (0 to
     ↪    50 units of time)
42   t_eval = np.linspace(t_span[0], t_span[1], n_step)   #
     ↪    Time points at which to evaluate the solution
43
44   # Target steady-state values for the system affected by
     ↪    Alzheimer (used in the genetic algorithm)
45   Cer_target = 6.0    # Firing rate of cerebellar neurons
     ↪    (Hz)
46   VTA_target = 5.0   # Firing rate of ventral tegmental
     ↪    area neurons (Hz)
47   mPFC_target = 4.0 # Firing rate of medial prefrontal
     ↪    cortex neurons (Hz)
48
49   mPFC_healthy = 7.0 # Firing rate of medial prefrontal
     ↪    cortex neurons for the healthy system (Hz). This
     ↪    data will be used for the plot
50
51   target_values = [Cer_target, VTA_target, mPFC_target] #
     ↪    Desired steady-state firing rates for Cer, VTA, and
     ↪    mPFC
52
53   # Genetic Algorithm (GA) parameters for optimizing the
     ↪    model's parameters
54   chromo_pop_num = 100   # Number of chromosomes in the
     ↪    population (population size)
55   mut_rate = 0.01         # Mutation rate for introducing
     ↪    small random changes to genes
```

```
56   surv_num = 1              # Number of survivors per
     ↪   generation (elitism)
57   gen_num = 30              # Total number of generations for
     ↪   the GA
58
59   # Initialize the first generation of genes (model
     ↪   parameters to be optimized)
60   first_gene_values_list = [a_ext, a_1, tau_Cer, b_ext,
     ↪   b_1, tau_VTA, c_ext, c_1, tau_mPFC]
61
62   # Solve the ODE system with the initial parameters to
     ↪   evaluate fitness before GA starts
63   sol = solve_ivp(model, t_span, x0,
     ↪   args=tuple(first_gene_values_list), t_eval=t_eval)
64
65   # Calculate the fitness based on the steady-state error
     ↪   (difference from target values)
66   predicted = np.array([sol.y[0, -1], sol.y[1, -1],
     ↪   sol.y[2, -1]])  # Predicted final values
67   errors = predicted - np.array(target_values)  #
     ↪   Difference from target
68   abs_perc_errors = np.abs((errors / predicted) * 100)  #
     ↪   Absolute percentage error
69   mean_ape = np.mean(abs_perc_errors)  # Mean absolute
     ↪   percentage error (MAPE)
70   start_fitness = (100 - mean_ape) / 100  # Fitness value
     ↪   (higher is better)
71
72   # Initialize the best fitness value and the selected
     ↪   population for the first generation
73   best_total_fitness = start_fitness  # Track the best
     ↪   fitness achieved so far
74   selected_pop = first_gene_values_list.copy()  # Set the
     ↪   initial best parameters as the selected population
75
76   # Store the fitness history to track progress across
     ↪   generations
77   total_fitness_history = []
78
79   # Genetic Algorithm main loop to evolve the population
     ↪   over multiple generations
80   for gen in range(gen_num):
81
82       # Create a new population by applying random
         ↪   mutations to the best solution from the
         ↪   previous generation
83       chromo_pop = []
84       for _ in range(chromo_pop_num):
85           # Apply a small mutation to each gene in the
             ↪   selected population
86           new_chromo = [gene + np.random.normal(0.0, gene
             ↪   * mut_rate) for gene in selected_pop]
87           chromo_pop.append(new_chromo)
```

```python
88
89        # Evaluate the fitness of each chromosome in the
          ↪  population
90        pop_total_fitness = []
91        for chromo in chromo_pop:
92            # Solve the ODE system with the chromosome's
              ↪  parameters to calculate fitness
93            sol = solve_ivp(model, t_span, x0,
              ↪  args=tuple(chromo), t_eval=t_eval)
94
95            # Calculate the steady-state values for this
              ↪  chromosome
96            predicted = np.array([sol.y[0, -1], sol.y[1,
              ↪  -1], sol.y[2, -1]])
97
98            # Compute fitness based on how closely the
              ↪  solution matches the target values (MAPE)
99            errors = predicted - np.array(target_values)
100           abs_perc_errors = np.abs((errors / predicted) *
              ↪  100)
101           mean_ape = np.mean(abs_perc_errors)
102           fitness = (100 - mean_ape) / 100  # Fitness
              ↪  value (higher is better)
103
104           pop_total_fitness.append(fitness)  # Append
              ↪  fitness of this chromosome to the
              ↪  population's fitness list
105
106       # Sort the population by fitness (highest first)
          ↪  and select the best chromosome
107       sorted_fitness_pop = sorted(zip(pop_total_fitness,
          ↪  chromo_pop), reverse=True)
108
109       # Update the best fitness and selected population
          ↪  for the next generation
110       best_total_fitness = sorted_fitness_pop[0][0]  #
          ↪  Best fitness in this generation
111       selected_pop = sorted_fitness_pop[0][1]  #
          ↪  Corresponding chromosome with the best fitness
112
113       # Store the best fitness value in the fitness
          ↪  history for future analysis
114       total_fitness_history.append(best_total_fitness)
115
116   # Plot the evolution of fitness over the generations
117   plt.figure(figsize=(10, 6))
118   plt.plot(total_fitness_history, label="Best Fitness
      ↪  Over Generations", marker='o')
119   plt.xlabel('Generation')  # Label for x-axis
120   plt.ylabel('Best Fitness')  # Label for y-axis
121   plt.title('Evolution of Fitness Over Generations')  #
      ↪  Title of the plot
122   plt.legend()  # Show legend on the plot
```

```
123    plt.grid(True)  # Enable grid for better readability
124    plt.show()
125
126    # Solve the ODE system using the best parameters found
       ↪   by the genetic algorithm
127    sol = solve_ivp(model, t_span, x0,
       ↪   args=tuple(selected_pop), t_eval=t_eval)
128
129    # Plot the results of the simulation with the evolved
       ↪   parameters
130    plt.figure(1)
131
132    plt.plot(sol.t, sol.y[0], 'b', linewidth=2, label='Cer')
133    plt.plot(sol.t, sol.y[1], 'g', linewidth=2, label='VTA')
134    plt.plot(sol.t, sol.y[2], 'r', linewidth=2,
       ↪   label='mPFC')
135
136    # Add horizontal lines for target values, limited to
       ↪   the time range
137    plt.plot([0, sol.t[-1]], [Cer_target, Cer_target],
       ↪   'b--', linewidth=1, label='Target Cer')
138    plt.plot([0, sol.t[-1]], [VTA_target, VTA_target],
       ↪   'g--', linewidth=1, label='Target VTA')
139    plt.plot([0, sol.t[-1]], [mPFC_target, mPFC_target],
       ↪   'r--', linewidth=1, label='Alzheimer mPFC')
140    plt.plot([0, sol.t[-1]], [mPFC_healthy, mPFC_healthy],
       ↪   'r--', linewidth=3, label='Healthy mPFC')
141
142
143    # Label the axes
144    plt.xlabel('Time')
145    plt.ylabel('Values')
146
147    # Add a legend outside the plot
148    plt.legend(loc='upper left', bbox_to_anchor=(1, 1))
149    plt.grid(True)
150
151    # Display the plot
152    plt.show()
153
154    # Display the parameters of the subject with mPFC
       ↪   deactivation due to Alzheimer's, as identified by
       ↪   the genetic algorithm
155    formatted_params = [f"{param:.2f}" for param in
       ↪   selected_pop]
156
157    # Print parameters in specified order
158    print("Parameters of the subject with mPFC deactivation
       ↪   due to Alzheimer's, as identified by the genetic
       ↪   algorithm:")
159    print(f"a_ext: {formatted_params[0]}")
160    print(f"a_1: {formatted_params[1]}")
161    print(f"tau_Cer: {formatted_params[2]}")
```

```
162   print(f"b_ext: {formatted_params[3]}")
163   print(f"b_1: {formatted_params[4]}")
164   print(f"tau_VTA: {formatted_params[5]}")
165   print(f"c_ext: {formatted_params[6]}")
166   print(f"c_1: {formatted_params[7]}")
167   print(f"tau_mPFC: {formatted_params[8]}")
```

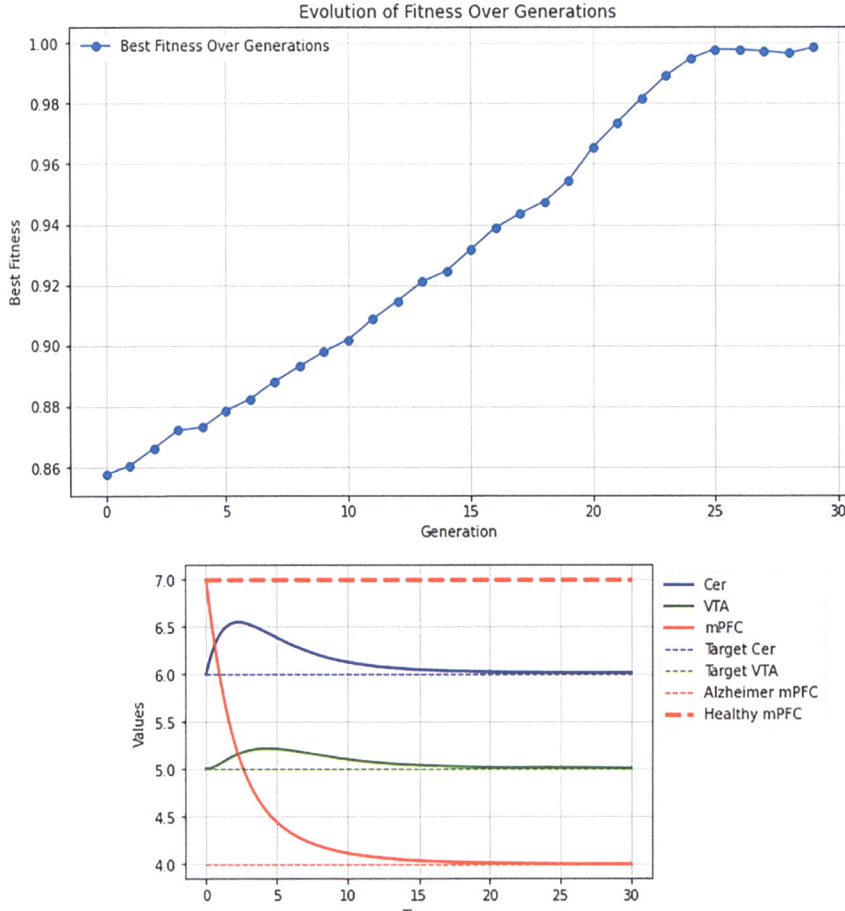

```
1   Parameters of the subject with mPFC deactivation due to
    ↪   Alzheimer's, as identified by the genetic algorithm:
2   a_ext: 1.93
3   a_1: 0.22
4   tau_Cer: 0.46
5   b_ext: 1.13
6   b_1: 0.27
```

```
7    tau_VTA: 0.55
8    c_ext: 0.55
9    c_1: 0.23
10   tau_mPFC: 0.42
```

The simulation results demonstrate that starting with the healthy subject's parameters and using the initial activities of the Cer, VTA, and mPFC, the genetic algorithm successfully identifies parameters that replicate the reduction in mPFC activity, as hypothesized in Alzheimer's disease. Simultaneously, the algorithm fine-tunes the parameters to preserve the activity levels of Cer and VTA. Comparing the parameters of the Alzheimer model with those of the healthy subject, the following differences are observed:

- **Changes in external region activity**: In the Alzheimer model, the parameter influencing Cer (a_{ext}) increases while the one affecting VTA (b_{ext}) is lower.
- **Changes in decay times**: The activity of Cer and VTA neurons decreases more slowly in the Alzheimer model, with smaller values for τ_{Cer} and τ_{VTA}. Conversely, the activity of mPFC neurons decreases faster, with a higher value for τ_{mPFC}.
- **Reduced excitation of VTA by Cer**: The parameter b_1, which represents the excitation of VTA by Cer, is lower in the Alzheimer model.
- **Reduced excitation of mPFC by VTA**: The parameter c_1, which represents the excitation of mPFC by VTA, is decreased in the Alzheimer model.

This section presented a simplified example with a few areas and a single simulated subject to illustrate the four-step method. However, the model can be expanded to simulate interactions among multiple areas across several subjects by varying the random seeds in the code. This approach facilitates statistical analysis and improves the robustness of the results. The final example in this chapter, which explores the role of monoamines in Parkinson's disease, illustrates a more complex procedure that integrates these considerations.

4.5   Other Examples

This section illustrates the application of the method in investigating learning mechanisms, the interaction between neural processes and body movement, and some neural processes underlying another neurodegenerative disorder, Parkinson's disease. In the first two examples, only steps two and three of the method are applied, as they involve straightforward cases that do not require block schema construction or parameter optimization. The third example, however, focuses on a complex model examining the role of monoamines in Parkinson's disease. This final case applies all four steps of the method, with detailed stability analysis and simulations of potential therapeutic interventions.

4.5.1 Simulating Learning Processes in the Basal Ganglia

Consider a scenario aimed at replicating learning mechanisms within the basal ganglia. The basal ganglia are a group of subcortical nuclei in the brain, primarily involved in motor control, procedural learning, and habit formation. They play a key role in regulating voluntary movements, reward-based learning, and decision-making processes by influencing cortical activity and modulating motor pathways. Dopamine, a neurotransmitter linked to the brain reward system, is essential in these learning processes, as it reinforces the association between actions and positive outcomes. Dopamine signals, particularly through reward prediction, enable the basal ganglia to adjust behavior and improve motor learning [67]. Their function is critical in conditions like Parkinson's disease, where dopamine depletion disrupts the basal ganglia ability to control movement and manage learning processes effectively [68]. The example below simulates dopaminergic modulation of striatal neuron activity (the striatum is a part of the basal ganglia), a key learning process in the basal ganglia. Given the simplicity of the task, generating a block diagram (step 1) is unnecessary. Instead, step 2 presents a simplified system of ODEs.

$$\frac{dS}{dt} = -aS + bD \tag{4.5.4}$$

$$\frac{dD}{dt} = -cD + \text{input} \tag{4.5.5}$$

The striatal activity (S) refers to the activity of neurons in the striatum; dopamine (D) indicates the level of dopamine, which modulates striatal activity; a, b, and c are parameters that define the dynamics of the system; "input" represents external stimuli or reward signals.

Now, proceed to implement this model using Python (step 3).

```python
import numpy as np
from scipy.integrate import solve_ivp
import matplotlib.pyplot as plt

# Define the differential equations for the basal
↪  ganglia model
def basal_ganglia_ode(t, y, a, b, c, input_function):
    """
    Computes the derivatives of striatal activity (S)
    ↪  and dopamine level (D)
    based on the current time t, state vector y, and
    ↪  model parameters.

    Parameters:
    - t: Current time
    - y: State vector [S, D]
    - a, b, c: Model parameters
    - input_function: Function that returns the external
    ↪  input based on time t
```

```python
16
17        Returns:
18        - List of derivatives [dS_dt, dD_dt]
19        """
20        S, D = y
21        dS_dt = -a * S + b * D
22        dD_dt = -c * D + input_function(t)
23        return [dS_dt, dD_dt]
24
25    # Define the input function (e.g., a time-varying reward
   ↪    signal)
26    def input_function(t):
27        """
28        Defines the external input to the system, which
   ↪        changes over time.
29
30        Parameters:
31        - t: Current time
32
33        Returns:
34        - Value of the input signal at time t
35        """
36        return np.sin(t) if t < 10 else 0
37
38    # Model parameters
39    a = 1.0  # Rate at which striatal activity decreases
40    b = 0.5  # Rate at which dopamine influences striatal
   ↪    activity
41    c = 1.0  # Rate at which dopamine decreases
42
43    # Initial conditions for the system
44    S0 = 0  # Initial striatal activity
45    D0 = 1  # Initial dopamine level
46
47    # Time span and evaluation points for the simulation
48    t_span = (0, 20)  # Time range from 0 to 20 units
49    t_eval = np.linspace(t_span[0], t_span[1], 400)  # 400
   ↪    time points for evaluation
50
51    # Solve the differential equations using the initial
   ↪    conditions and parameters
52    solution = solve_ivp(basal_ganglia_ode, t_span, [S0,
   ↪    D0], args=(a, b, c, input_function), t_eval=t_eval)
53
54    # Plotting the simulation results
55    plt.figure(figsize=(10, 5))
56    plt.plot(solution.t, solution.y[0], label='Striatal
   ↪    Activity (S)')
57    plt.plot(solution.t, solution.y[1], label='Dopamine
   ↪    Level (D)')
58    plt.title('Simulation of Learning Mechanisms in the
   ↪    Basal Ganglia')
59    plt.xlabel('Time')
```

```
60   plt.ylabel('Activity')
61   plt.legend()
62   plt.grid(True)
63   plt.show()
```

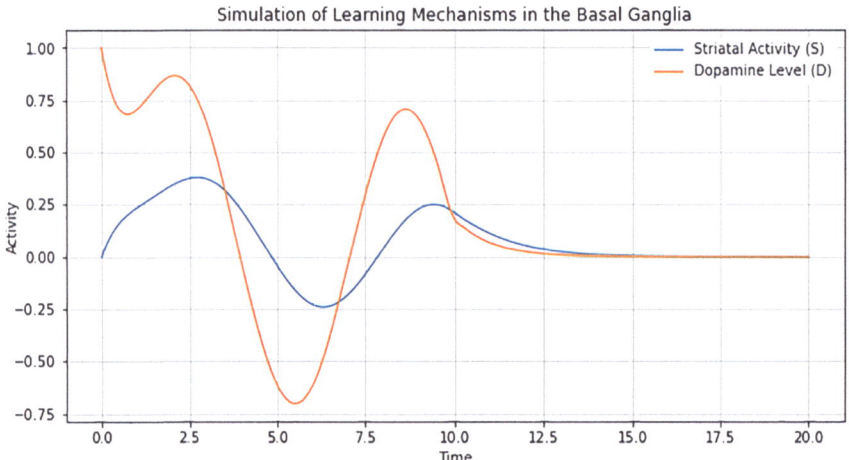

The rate of change in striatal activity, denoted as $\frac{dS}{dt}$, reflects how striatal activity evolves over time. This rate is influenced by the intrinsic decay of striatal activity and the modulation provided by dopamine levels. Similarly, the rate of change in dopamine levels, $\frac{dD}{dt}$, accounts for the natural decay of dopamine over time, while also being affected by external stimuli. The input_function models an external stimulus or reward signal that impacts dopamine levels. In this model, the function is represented as a sine wave for the first 10 time units and then ceases, simulating a varying external input. Constants a, b, and c are parameters that govern the rate at which striatal activity and dopamine levels change. These constants are crucial for determining the dynamics of the model. Initial conditions specify the starting values for striatal activity and dopamine levels, while the time span of the simulation is set to observe the behavior over a defined period.

The simulation results are visualized through plotting, which shows how striatal activity and dopamine levels change over time. This simplified model offers insights into how the basal ganglia might process learning by examining the interaction between striatal activity and dopamine modulation. More complex models could incorporate additional neural populations and interactions to more accurately reflect the physiology of the basal ganglia.

4.5.2 Simulating Basic Brain-Body Interactions

This example utilizes an ODEs system to model the primary motor cortex and its connection to a robotic arm with two degrees of freedom (2DOF). The `solve_ivp` function will be employed to solve the ODEs and simulate the movement of the robotic arm as controlled by the motor cortex. Given the simplicity of the model a block diagram is unnecessary; instead, the focus shifts directly to formulating the ODEs system (step 2). The following equations describe the dynamics of the motor cortex, modeling the relationship between its state variables and the output.

$$\frac{d\theta_1}{dt} = -\frac{\theta_1}{\tau} \tag{4.5.6}$$

$$\frac{d\theta_2}{dt} = -\frac{\theta_2}{\tau} \tag{4.5.7}$$

$$\frac{dmotor\_output}{dt} = \alpha(1 - motor\_output) \tag{4.5.8}$$

where:

- θ_1 and θ_2 represent the state variables of the motor cortex, which can be thought of as neural activity levels.
- τ is the time constant for the motor cortex dynamics, reflecting how quickly the system responds to changes.
- $motor\_output$ is the output signal from the motor cortex that influences motor commands.
- α is the gain factor that determines how rapidly the motor output approaches its maximum value.

These equations capture the continuous dynamics of the motor cortex, where the state variables decay over time based on their respective time constants, and the motor output is driven towards a desired level by the factor α. The expression $(1 - motor\_output)$ can play a significant role in the dynamics of robotic arm control, particularly when linking motor cortex activity to arm movements. This formulation can be understood through several interconnected concepts. First, the expression may represent an inverse relationship between the motor output and the control exerted on the robotic arm. In certain scenarios, as the motor output increases, one might want the resulting effect on the arm dynamics to decrease. This could be essential for preventing overly aggressive movements, ensuring that the arm actions remain smooth and controlled. By employing $(1 - motor\_output)$, the model captures this dynamic by reducing the influence on the arm movements as the motor output becomes more pronounced. Another important aspect is the concept of normalization. In control systems, outputs often need to operate within a specific range. If the motor output varies between 0 and 1, using $(1 - motor\_output)$ can help to maintain the values within the desired limits. This normalization allows for a more consistent and predictable behavior of the robotic arm, as it ensures that the parameters remain within a suitable range for effective control. Moreover, the expression can be seen as a way to introduce damping or resistance into the system. In many real-world applications, higher control signals may not always lead to proportionally higher movements. By incorporating (1 − motor_output), the model can simulate a damping effect, where stronger commands from the motor cortex lead to a more nuanced response in the arm dynamics. This is particularly relevant when precision is required, as it helps in moderating the response to avoid jerky or erratic movements. Finally, (1 − motor_output) can facilitate a feedback mechanism in the control system. In feedback control, the goal is often to minimize the difference between desired and actual states. By representing the motor output in this manner, the expression can reflect the system ability to adjust its movements based on the current output from the motor cortex. For example, when the motor output approaches 1, the corresponding (1 − motor_output) value decreases, which could signal to the system that less correction is needed, thereby allowing for more precise movement adjustments. Thus, the choice of using (1 − motor_output) in the dynamics of the robotic arm serves multiple purposes, including modeling inverse relationships, normalizing control signals, introducing damping effects, and enabling feedback mechanisms. Each of these elements contributes to a more sophisticated and realistic representation of how motor cortex activity influences robotic arm dynamics, ensuring that the arm operates effectively and smoothly in response to its intended commands.

The ODEs presented above are implemented in the Python code (step 3) using the *motor_cortex* function. In the Python implementation, the derivatives are computed and returned as an array, allowing the *solve_ivp* function to numerically integrate these equations over time. The output of the motor cortex, captured in the variable *motor_output*, is subsequently utilized in the *robotic_arm* function to influence the dynamics of the robotic arm, demonstrating the connection between the motor cortex and arm movements.

```
1   from scipy.integrate import solve_ivp  # Import the
    ↪  solve_ivp function for solving ODEs
2   import numpy as np  # Import NumPy for numerical
    ↪  operations
3   import matplotlib.pyplot as plt  # Import Matplotlib
    ↪  for plotting
4   from scipy.interpolate import interp1d  # Import
    ↪  interp1d for interpolation of motor outputs
5
6   # Define constants for the robotic arm dynamics
7   m1 = 1.0  # Mass of link 1
8   m2 = 1.0  # Mass of link 2
9   l1 = 1.0  # Length of link 1
10  l2 = 1.0  # Length of link 2
11  g = 9.81  # Acceleration due to gravity
12
13  # Define parameters for motor cortex dynamics
14  tau = 1.0        # Time constant for the motor cortex
    ↪  dynamics
```

```python
15    alpha = 1.0       # Gain factor for motor output influence
16
17    # Define the function representing the dynamics of the
      ↪  motor cortex
18    def motor_cortex(t, y):
19        """
20        Dynamics of the motor cortex.
21
22        Parameters:
23            t: float
24                Time variable.
25            y: array_like
26                Array containing the state variables
                 ↪  [theta1, theta2, motor_output].
27
28        Returns:
29            dydt: array_like
30                Array containing the derivatives of the
                 ↪  state variables [dtheta1/dt,
                 ↪  dtheta2/dt, d(motor_output)/dt].
31        """
32
33        theta1, theta2, motor_output = y  # Unpack state
          ↪  variables
34
35        # ODEs for the motor cortex
36        dydt = [
37            -theta1 / tau,        # Rate of change of theta1
              ↪  (placeholder dynamics)
38            -theta2 / tau,        # Rate of change of theta2
              ↪  (placeholder dynamics)
39            alpha * (1 - motor_output)  # Motor cortex
              ↪  output influences itself (encouraging
              ↪  growth towards a target)
40        ]
41        return dydt  # Return the computed derivatives
42
43    # Define the function representing the dynamics of the
      ↪  robotic arm
44    def robotic_arm(t, y, motor_output_func):
45        """
46        Dynamics of the robotic arm.
47
48        Parameters:
49            t: float
50                Time variable.
51            y: array_like
52                Array containing the state variables
                 ↪  [theta1, theta2, theta1_dot,
                 ↪  theta2_dot].
53            motor_output_func: callable
54                A function that returns the output of the
                 ↪  motor cortex at time t.
```

```
55
56      Returns:
57          dydt: array_like
58              Array containing the derivatives of the
         ↪    state variables [dtheta1/dt,
         ↪    dtheta2/dt, d(theta1_dot)/dt,
         ↪    d(theta2_dot)/dt].
59      """
60
61      theta1, theta2, theta1_dot, theta2_dot = y  #
         ↪    Unpack state variables
62
63      # Get the motor output at the current time step
64      motor_output = motor_output_func(t)  # Call the
         ↪    function to get motor output
65
66      # Influence of motor output on the arm dynamics
67      motor_effect_theta1 = motor_output * 0.5  # Scaling
         ↪    the influence on joint 1 (modify as needed)
68      motor_effect_theta2 = motor_output * 0.5  # Scaling
         ↪    the influence on joint 2 (modify as needed)
69
70      # Equations of motion for the robotic arm
71      theta1_ddot = (m2 * l1 * theta1_dot ** 2 *
         ↪    np.sin(theta2 - theta1) +
72                       m2 * g * np.sin(theta2) *
                          ↪    np.cos(theta2 - theta1) +
73                       m2 * l2 * theta2_dot ** 2 *
                          ↪    np.sin(theta2 - theta1) -
74                       (m1 + m2) * g * np.sin(theta1)) /
                          ↪    (l1 * (m1 + m2 * (1 -
                          ↪    np.cos(theta2 - theta1) ** 2)))
75
76      theta2_ddot = (-m2 * l2 * theta2_dot ** 2 *
         ↪    np.sin(theta2 - theta1) +
77                       (m1 + m2) * (g * np.sin(theta1) *
                          ↪    np.cos(theta2 - theta1) - l1 *
                          ↪    theta1_dot ** 2 * np.sin(theta2
                          ↪    - theta1) -
78                                    g * np.sin(theta2))) /
                                      ↪    (l2 * (m1 + m2 * (1
                                      ↪    - np.cos(theta2 -
                                      ↪    theta1) ** 2)))
79
80      # Integrate the motor_output into the dynamic
         ↪    equations
81      theta1_ddot += motor_effect_theta1  # Modify
         ↪    theta1's acceleration based on motor cortex
         ↪    output
82      theta2_ddot += motor_effect_theta2  # Modify
         ↪    theta2's acceleration based on motor cortex
         ↪    output
83
```

```
84      dydt = [theta1_dot, theta2_dot, theta1_ddot,
         ↪  theta2_ddot]  # Pack the derivatives
85      return dydt  # Return the computed derivatives
86
87  # Initial conditions for the simulation
88  theta1_0 = 0.0  # Initial angle of joint 1 (in radians)
89  theta2_0 = 0.0  # Initial angle of joint 2 (in radians)
90  theta1_dot_0 = 0.0  # Initial angular velocity of joint
     ↪  1
91  theta2_dot_0 = 0.0  # Initial angular velocity of joint
     ↪  2
92  motor_output_0 = 0.5  # Initial motor cortex output
     ↪  (influence)
93
94  # Time span for the simulation
95  t_span = (0, 10)  # From t=0 to t=10 seconds
96  t_eval = np.linspace(t_span[0], t_span[1], 100)  # 100
     ↪  time points for evaluation
97
98  # Solve the ODE system for the motor cortex
99  sol_motor = solve_ivp(motor_cortex, t_span, [theta1_0,
     ↪  theta2_0, motor_output_0], t_eval=t_eval)
100
101 # Create a function to interpolate the motor output
     ↪  over time
102 motor_output_func = interp1d(sol_motor.t,
     ↪  sol_motor.y[2], fill_value="extrapolate")  #
     ↪  Interpolation function for motor output
103
104 # The motor cortex output influences the robotic arm's
     ↪  movement.
105 # Solve the ODE system for the robotic arm, using the
     ↪  interpolated motor cortex output function
106 sol_robotic_arm = solve_ivp(robotic_arm, t_span,
     ↪  [theta1_0, theta2_0, theta1_dot_0, theta2_dot_0],
107                             args=(motor_output_func,),
                                 ↪  t_eval=t_eval)
108
109 # Import matplotlib for plotting
110 import matplotlib.pyplot as plt
111
112 # Create a figure for plotting the results
113 plt.figure(figsize=(10, 5))
114
115 # Plot motor cortex output
116 plt.subplot(2, 1, 1)  # Create a subplot for motor
     ↪  cortex output
117 plt.plot(sol_motor.t, sol_motor.y[2], 'r-',
     ↪  linewidth=2)  # Plot motor output over time
118 plt.title('Motor Cortex Output')  # Title for the motor
     ↪  output plot
119 plt.xlabel('Time (s)')  # X-axis label
120 plt.ylabel('Motor Output')  # Y-axis label
```

```
121   plt.grid()  # Add grid for better readability
122
123   # Plot robotic arm movement
124   plt.subplot(2, 1, 2)  # Create a subplot for robotic
      ↪   arm movement
125   plt.plot(sol_robotic_arm.t, sol_robotic_arm.y[0], 'b-',
      ↪   label='Joint 1 (Theta1)')  # Plot joint 1 angle
126   plt.plot(sol_robotic_arm.t, sol_robotic_arm.y[1], 'g-',
      ↪   label='Joint 2 (Theta2)')  # Plot joint 2 angle
127   plt.title('Robotic Arm Movement')  # Title for the
      ↪   robotic arm movement plot
128   plt.xlabel('Time (s)')  # X-axis label
129   plt.ylabel('Angle (radians)')  # Y-axis label
130   plt.legend()  # Show legend for the joint labels
131   plt.grid()  # Add grid for better readability
132
133   # Adjust layout and display the plots
134   plt.tight_layout()  # Adjust subplot parameters for a
      ↪   clean layout
135   plt.show()  # Display the plots
```

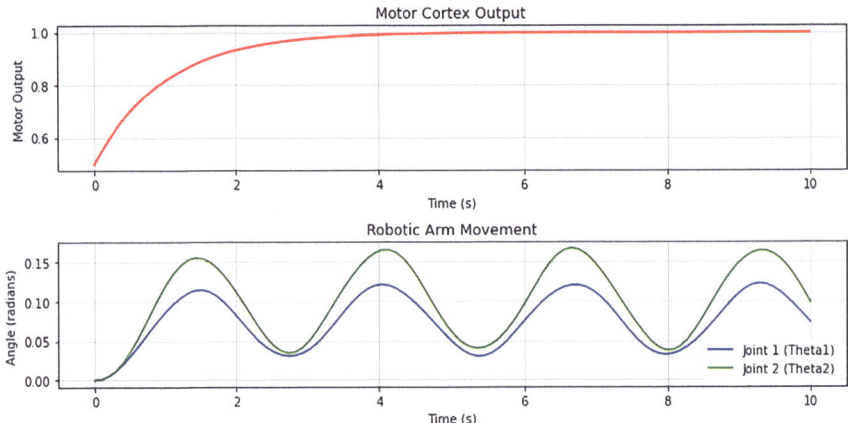

The following code generates an animation that visually represents the movements of a robotic arm controlled by the motor output signals from the motor cortex. This dynamic visualization allows for an interactive observation of how the motor cortex influences the robotic arm joint movements over time (Fig. 4.3).

Fig. 4.3 Snapshot of the robotic arm position at frame 50 during the simulation, illustrating the joint configuration and end-effector location

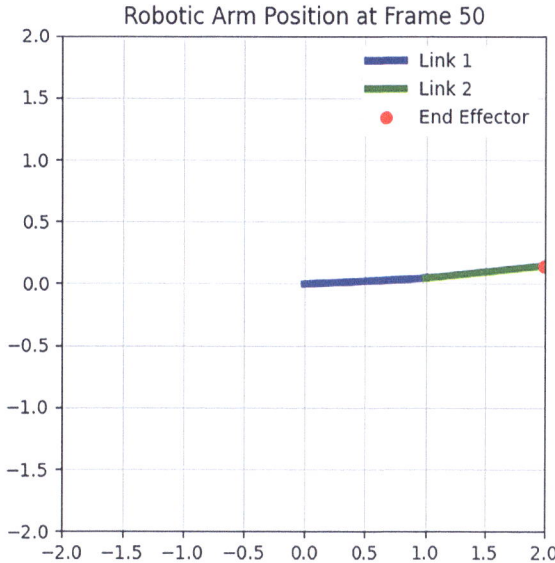

```
1   import numpy as np
2   import matplotlib.pyplot as plt
3   from matplotlib.animation import FuncAnimation
4
5   # Ensure that the following arrays are available from prior
    ↪    computations:
6   # sol_motor: output from the motor cortex ODE, containing the
    ↪    motor output signal
7   # sol_robotic_arm: output from the robotic arm ODE,
    ↪    containing joint angle data
8
9   # Define a range of time steps for the animation (0 to 10
    ↪    seconds, with 100 intervals)
10  t_eval = np.linspace(0, 10, 100)
11  motor_output = sol_motor.y[2]   # Extract the motor output
    ↪    signal from the motor cortex solution
12  theta1 = sol_robotic_arm.y[0]   # Extract the angle of the
    ↪    first joint from the robotic arm solution
13  theta2 = sol_robotic_arm.y[1]   # Extract the angle of the
    ↪    second joint from the robotic arm solution
14
15  # Set the lengths of the two segments of the 2DOF robotic arm
16  l1 = 1.0   # Length of the first arm segment
17  l2 = 1.0   # Length of the second arm segment
18
19  # Function to compute the end effector position based on
    ↪    joint angles
20  def compute_end_effector(theta1, theta2):
21      # Calculate the x and y coordinates of the end effector
    ↪    using forward kinematics
```

```python
22        x = l1 * np.cos(theta1) + l2 * np.cos(theta1 + theta2)
23        y = l1 * np.sin(theta1) + l2 * np.sin(theta1 + theta2)
24        return x, y
25
26    # Create a figure for the animation
27    fig, ax = plt.subplots()
28    # Set the limits of the plot to accommodate the arm's movement
29    ax.set_xlim(-2, 2)
30    ax.set_ylim(-2, 2)
31    ax.set_aspect('equal')   # Ensure equal scaling of x and y axes
32    ax.grid()   # Add a grid for better visualization of the
      ↪   movement
33
34    # Initialize the line objects for the arm segments and end
      ↪   effector
35    line1, = ax.plot([], [], 'b-', lw=4, label='Link 1')   # First
      ↪   arm segment (blue)
36    line2, = ax.plot([], [], 'g-', lw=4, label='Link 2')   #
      ↪   Second arm segment (green)
37    end_effector, = ax.plot([], [], 'ro', label='End Effector')
      ↪   # End effector (red)
38
39    # Add a legend to identify the components
40    ax.legend()
41
42    # Initialization function for the animation, setting up empty
      ↪   data for each frame
43    def init():
44        line1.set_data([], [])   # Reset the first line data
45        line2.set_data([], [])   # Reset the second line data
46        end_effector.set_data([], [])   # Reset the end effector
          ↪   data
47        return line1, line2, end_effector   # Return updated
          ↪   objects
48
49    # Update function for the animation, which updates the arm's
      ↪   position for each frame
50    def update(frame):
51        # Calculate the positions of the joints based on the
          ↪   current angles
52        x1 = l1 * np.cos(theta1[frame])   # x-coordinate of the
          ↪   first joint
53        y1 = l1 * np.sin(theta1[frame])   # y-coordinate of the
          ↪   first joint
54        # Compute the end effector's position using the current
          ↪   joint angles
55        x2, y2 = compute_end_effector(theta1[frame],
          ↪   theta2[frame])
56
57        # Update the data for the arm segments
58        line1.set_data([0, x1], [0, y1])   # Update Link 1
59        line2.set_data([x1, x2], [y1, y2])   # Update Link 2
60        end_effector.set_data(x2, y2)   # Update the position of
          ↪   the end effector
```

```
61        return line1, line2, end_effector  # Return updated
    ↪   objects
62
63    # Create the animation by repeatedly calling the update
    ↪   function for each frame
64    ani = FuncAnimation(fig, update, frames=len(t_eval),
    ↪   init_func=init, blit=True, interval=50)
65
66    # Display the animated plot
67    plt.show()
```

```
ı    <IPython.core.display.Javascript object>
```

```
ı    <IPython.core.display.HTML object>
```

The function `interp1d` from the `scipy.interpolate` module is employed to facilitate the interpolation of motor outputs. This approach is essential for ensuring that the motor cortex output data, which is sampled at discrete time intervals, can be effectively applied to the dynamics of the robotic arm. By interpolating these outputs, a continuous function is created that estimates motor outputs at any given point in time, thereby enhancing the responsiveness and accuracy of the robotic arm movements. Interpolation allows for bridging the gaps between the sampled motor outputs, enabling smoother transitions and more precise control over the robotic arm joint angles. As a result, the arm can react in real-time to the evolving motor signals, reflecting a more realistic representation of how the motor cortex influences arm dynamics. This technique is particularly important in dynamic systems, where timely adjustments based on continuous inputs are crucial for achieving desired outcomes.

The results of the simulations, as depicted in the plots, illustrate the interaction between the motor cortex output and the dynamics of the robotic arm. In the first subplot, titled "Motor Cortex Output," the red line represents the motor output over time. This output reflects how the motor cortex modulates signals that influence movement. The fluctuations in the motor output indicate varying levels of activation, which are critical for coordinating motor commands. In the second subplot, titled "Robotic Arm Movement," the blue and green lines represent the angles of Joint 1 and Joint 2 of the robotic arm, respectively. The trajectories of these angles demonstrate how the robotic arm responds to the motor cortex output. The changes in joint angles correspond to the variations in motor output, illustrating how the motor cortex drives the arm movements. The plots effectively demonstrate the relationship between the motor cortex activity and the resulting joint dynamics of the robotic arm. The ability to visualize this relationship provides valuable insights into the control mechanisms underlying motor functions and the potential applications of this model in robotics and neurophysiology.

Increased motor output results in oscillatory joint movements due to several factors. Higher motor output activates motor pathways more strongly, leading to larger angular displacements and oscillations. The robotic arm nonlinear dynamics often cause oscillations when subjected to strong forces. Additionally, feedback loops between joints can create over-corrections, contributing to this behavior. The arm inertia and insufficient damping may also prevent effective stabilization, allowing sustained oscillations. Thus, the interplay of motor output, system dynamics, feedback interactions, and physical properties drives the observed oscillatory movements.

In this example, the focus is solely on using the `motor_output` from the motor cortex to influence the dynamics of the robotic arm. This output serves as a control signal that affects the acceleration of the arm joints. However, to utilize the θ outputs of the motor cortex more explicitly, such as using them as desired joint angles, the robotic arm dynamics could be modified accordingly. By referencing these angles directly, the arm trajectory can be adjusted more precisely. Implementing a feedback control system could further enhance this process by continuously comparing the desired joint angles with the actual angles of the arm. This comparison allows for real-time adjustments, refining the arm movement to achieve more accurate and responsive behavior, similar to how biological systems operate in response to neural commands.

4.5.3 Simulating the Effect of Monoamine Depletion on Parkinson's Disease

In [3], the authors propose a dynamic model designed to explore the interactions of three key brain monoamines in Parkinson's disease: dopamine (DA), noradrenaline (NA), and serotonin (5-HT). The study follows exactly the four step method approach: a system-level model is identified, translated in a system of differential equations, which is then implemented and matched to the available experimental data. DA plays a central role in movement control, reward, and motivation. Its depletion is a hallmark of Parkinson's disease, leading to the motor symptoms characteristic of the condition. NA affects attention, arousal, and stress responses, and 5-HT is crucial for mood regulation, sleep, and cognition. Despite their significant roles and the complex interactions between these three monoamines, most research on Parkinson's disease has focused primarily on dopamine. The dynamic interplay between DA, NA, and 5-HT remains mostly underexplored, even though these neurotransmitters profoundly influence each other and may collectively contribute to the disease broader range of symptoms.

The average activation frequency of neurons from the brain regions primarily responsible for releasing these monoamines-such as the globus pallidus (GP), striatal spiny neurons of type D1 and D2 (StrD1, StrD2), substantia nigra pars compacta and ventral tegmental area complex (SNcVTA), dorsal raphe nucleus (DRN), and locus coeruleus (LC)-were used as proxy measurements for monoamine levels. These frequencies were chosen as representative measurements because they were the only reliable empirical data from multiple sources. The diagram in Fig. 4.4 illustrates the hypothesized interactions among the areas under study and the monoamines. This conceptual interaction schema translates into the following system of differential equations:

$$\dot{GP} = -\frac{1}{\tau_{GP}}GP - \alpha_{GP}^{StrD1}StrD1 - \alpha_{GP}^{StrD2}StrD2 + \alpha_{GP}^{DRN}DRN + \alpha_{GP}^{ext} \quad (4.5.9)$$

$$\dot{StrD1} = -\frac{1}{\tau_{StrD1}}StrD1 + \alpha_{StrD1}^{SNcVTA}SNcVTA + \alpha_{StrD1}^{DRN}DRN + \alpha_{StrD1}^{ext} \quad (4.5.10)$$

$$\dot{StrD2} = -\frac{1}{\tau_{StrD2}}StrD2 - \alpha_{StrD2}^{SNcVTA}SNcVTA + \alpha_{StrD2}^{DRN}DRN + \alpha_{StrD2}^{ext} \quad (4.5.11)$$

$$\dot{SNcVTA} = -\frac{1}{\tau_{SNcVTA}}SNcVTA - \alpha_{SNcVTA}^{DRN}DRN - \alpha_{SNcVTA}^{LC}LC$$
$$+ \beta_{SNcVTA}^{LC}LC^2 + \alpha_{SNcVTA}^{ext} \quad (4.5.12)$$

$$\dot{DRN} = -\frac{1}{\tau_{DRN}}DRN - \alpha_{DRN}^{SNcVTA}SNcVTA + \alpha_{DRN}^{LC}LC + \alpha_{DRN}^{ext} \quad (4.5.13)$$

$$\dot{LC} = -\frac{1}{\tau_{LC}}LC + \alpha_{LC}^{SNcVTA}SNcVTA - \alpha_{LC}^{DRN}DRN + \alpha_{LC}^{ext} \quad (4.5.14)$$

where the abbreviated notation $\dot{x}$ stands for $\frac{dx}{dt}$ and:

- the time constants τ_x are all positive and refer to a dampening term which brings back the activity of each area to its resting activation level in the absence of external stimulation
- the parameters α represent the linear components of the system, are all positive and follow the notation: α_{to}^{from}; α_x^{ext} are synthetic terms that implicitly account for the rest activation of each area and other external stimuli which are not part of the modelled circuit; β is also positive and follow the same notation β_{to}^{from}, but account for nonlinear effects;
- the ratios of monoaminic projections from an area to its targets are assumed to be constant.

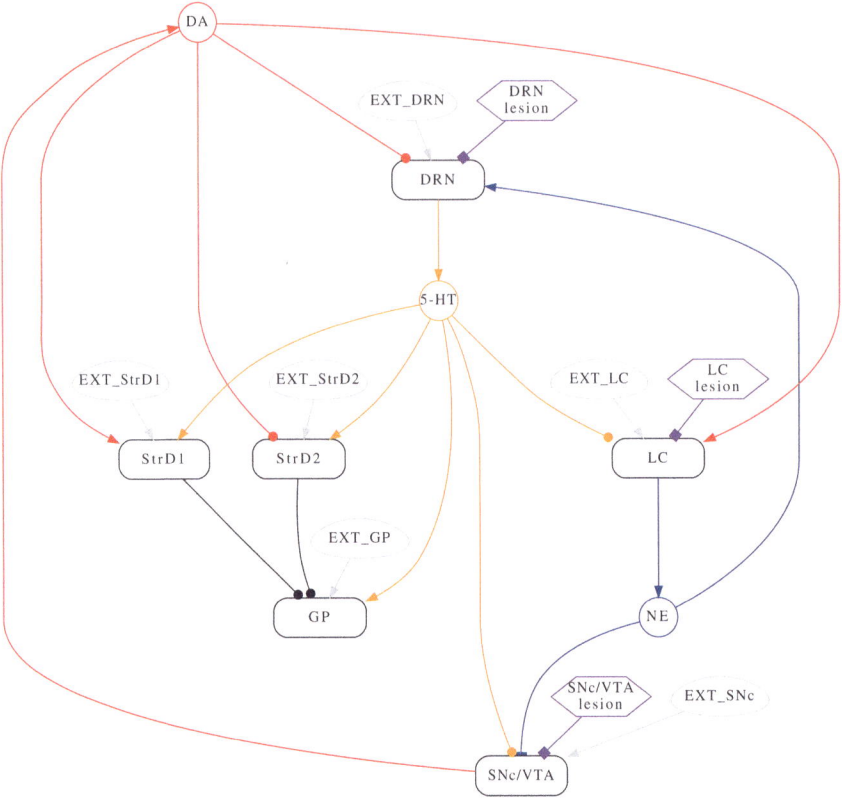

Fig. 4.4 Conceptual model schema. The average activation frequencies of six brain areas are modeled (rounded rectangles); some interactions are modulated by monoamines (circles). Arrows represent positive (excitatory) effects while circles represent negative (inhibitory) effects. Noradrenaline has a nonlinear (both excitatory and inhibitory) effect on SNcVTA which is indicated by a bar. Each area has a corresponding stimulus (ovals) which represents self-activation as well all any other stimulus the area might receive from the rest of the brain which is not modeled. Lastly, hexagons serve as indicators for identifying the areas influenced by the administration of specific drugs. Reproduced with permission from [3]

The system can be expressed in vectorial form by defining the status vector **y**:

$$\mathbf{y} = (GP, StrD1, StrD2, SNcVTA, DRN, LC)^T \in \mathbb{R}^s \qquad (4.5.15)$$

Since the non-linear part is limited to quadratic terms, the system defining function can be represent as a sum of three components:

$$\dot{\mathbf{y}}(t) = \mathbf{f}(\mathbf{y}(t)) = A\mathbf{y}(t) + C(\mathbf{y}(t) \circ \mathbf{y}(t)) + \mathbf{b} \qquad (4.5.16)$$

where $a_{ij} = 0$ if the corresponding α_i^j is not defined and likewise $b_i = 0$ if α_i^{ext} is not defined, and also $c_{ij} = \beta_i^j$ with $c_{ij} = 0$ if the corresponding β_i^j coefficient is not defined; "∘" indicates the element-wise vector product (or Hadamard).

Each component of the status vector **y** directly represents the average activation frequency of a brain area, and is therefore expressed in Hz. The derivative term in each equation of system (4.5.9)–(4.5.14) are all derivatives with respect to time of a frequency, hence they are all expressed in Hz/s (or $1/s^2$). Consequently, the external stimulus parameters α^{ext} must also be expressed in Hz/s, while the remaining α parameters must be 1/s, hence Hz.

The second order term parameters β are instead pure numbers, since $Hz^2 = 1/s^2 = Hz/s$. Finally, all time constants τ are naturally expressed in seconds.

The model is supposed to be able to reproduce multiple states of the brain in different conditions:

- Healthy (or SHAM)
- Serotonin deficiency
- Dopamine deficiency
- Noradrenaline deficiency
- Combined Dopamine and Noradrenaline deficiency
- Combined Dopamine and Serotonin deficiency.

In particular, the authors were not interested in the dynamical behaviour of the system, but only on the steady state stable asymptotic solution, which would identify a particular condition. The base hypothesis is that if the model is correct, while the circuit remains unchanged (hence, the equations remain the same), different sets (or subsets) of parameters should be enough to model correctly all the situations listed above.

Each element within the status vector (4.5.15) corresponds to the average activation frequency of the corresponding brain region. This, in turn, serves as an indirect indicator of the production and projection of monoamines to the affected areas. We posit that a depletion in monoamines results from the death or temporary incapacitation of a portion of neurons within a given area, which is directly manifested as a decrease in the average activation frequency of that area. Consequently, when the SNcVTA is lesioned, we observe a reduction in DA levels; lesioning the DRN results in a decrease in 5-HT, and lesioning the LC leads to a reduction in NA levels.

Each equation of the model is composed by three conceptual blocks: a damping term, a constant stimulus and a reaction to projections from other areas. The constant stimulus represents external and internal activation sources that are not directly accounted for in this model. Together with the damping term, the constant stimulus accounts for the resting behaviour of the area: the area will stabilize to its rest activation frequency. In absence of reaction terms, each equation has an equilibrium point:

$$y'(t) = -\frac{1}{\tau}y(t) + k, \quad y'(t) \equiv 0 \Rightarrow y(t) = k\tau. \qquad (4.5.17)$$

The time constants τ are derived from literature and we assume them to be typical values for the specific kind of neuron found in an area; we therefore assume that they are not altered by the lesion. It is, however, reasonable to expect that the sensitivity of lesioned area to internal and external stimuli will change in such a way that the average activation frequency changes to the levels which have been experimentally measured.

We can now define multiple versions of the same model, which differ from the healthy model only for the constant term and reaction coefficients of the lesioned area. For example, suppose a healthy subject is modeled using model (4.5.16) by the coefficients held in A, C, $\mathbf{b}$. Having received a dopaminergic lesion (hence, SNcVTA neurons are malfunctioning), the subject will now be modeled by the same equations of model (4.5.16) but this time with coefficients A_{LDA}, C_{LDA}, $\mathbf{b}_{LDA}$, which differ by A, C, $\mathbf{b}$ only by the values corresponding to the parameters of the equation for SNcVTA, namely α_{SNcVTA}^{DRN}, α_{SNcVTA}^{LC}, β_{SNcVTA}^{LC}, α_{SNcVTA}^{ext}. Likewise, when the subject also receives a serotonergic lesion, there will be a third set of parameters $A_{LDA+L5HT}$, $C_{LDA+L5HT}$, $\mathbf{b}_{LDA+L5HT}$ which again differ from A_{LDA}, C_{LDA}, $\mathbf{b}_{LDA}$ only by the parameters corresponding to the equation for DRN, and so on.

A single subject is therefore represented by multiple versions of the parameters matrices A, C, $\mathbf{b}$, each set corresponding to one particular state: healthy (also called SHAM), LDA, L5HT, LNE when only one of the lesions is applied, LDA+L5HT, LDA+LNE when lesions are combined, and so on.

4.5.3.1   Stability

A system like (4.5.16) possesses at least one equilibrium point denoted as $\bar{\mathbf{y}}$, satisfying the following condition:

$$0 = A\bar{\mathbf{y}} + C(\bar{\mathbf{y}} \circ \bar{\mathbf{y}}) + \mathbf{b} \tag{4.5.18}$$

The original system (4.5.16) can be expressed as:

$$\dot{\mathbf{y}}(t) = A\mathbf{y}(t) + C(\mathbf{y}(t) \circ \mathbf{y}(t)) + \mathbf{b} = f(t, \mathbf{y}) \tag{4.5.19}$$

This system can be translated to have an equilibrium point at the origin. Let $\bar{\mathbf{y}}$ represent the equilibrium point, such that:

$$0 = A\bar{\mathbf{y}} + C(\bar{\mathbf{y}} \circ \bar{\mathbf{y}}) + \mathbf{b} = f(t, \bar{\mathbf{y}}) \tag{4.5.20}$$

Substituting $\mathbf{y} + \bar{\mathbf{y}}$ into the system yields:

$$
\begin{aligned}
(\mathbf{y} + \bar{\mathbf{y}})' &= \dot{\mathbf{y}} \\
&= A(\mathbf{y} + \bar{\mathbf{y}}) + C((\mathbf{y} + \bar{\mathbf{y}}) \circ (\mathbf{y} + \bar{\mathbf{y}})) + \mathbf{b} \\
&= A\mathbf{y} + A\bar{\mathbf{y}} + C(\mathbf{y} \circ \mathbf{y}) + 2C(\mathbf{y} \circ \bar{\mathbf{y}}) + C(\bar{\mathbf{y}} \circ \bar{\mathbf{y}}) + \mathbf{b} \\
&= A\mathbf{y} + C(\mathbf{y} \circ \mathbf{y}) + 2C(\mathbf{y} \circ \bar{\mathbf{y}}) \text{ since } (4.5.20) \text{ holds.} \quad (4.5.21)
\end{aligned}
$$

Now, it is clear that $\mathbf{y} = \mathbf{0}$ represents an equilibrium point, as both A and C are linear combinations, and the element-wise product with the zero vector results in zero.

Let $D_{\bar{\mathbf{y}}} = \operatorname{diag}(\bar{\mathbf{y}})$. The term $\mathbf{y} \circ \bar{\mathbf{y}}$ can be rewritten as $D_{\bar{\mathbf{y}}}\mathbf{y}$. Thus, the equation simplifies to:

$$
\dot{\mathbf{y}} = A\mathbf{y} + C(\mathbf{y} \circ \mathbf{y}) + 2CD_{\bar{\mathbf{y}}}\mathbf{y} = (\tilde{A})\mathbf{y} + C(\mathbf{y} \circ \mathbf{y}) \quad (4.5.22)
$$

Here, $\tilde{A} = A + 2CD_{\bar{\mathbf{y}}}$. The system can then be represented as a sum of a linear and a nonlinear term:

$$
\dot{\mathbf{y}}(t) = \tilde{A}\mathbf{y}(t) + \mathbf{g}(\mathbf{y}(t)) \quad (4.5.23)
$$

where $\mathbf{g}(\mathbf{y}) = C(\bar{\mathbf{y}} \circ \bar{\mathbf{y}})$. In this formulation, the matrices A, C, and $D_{\bar{\mathbf{y}}}$ remain constant with respect to time in (4.5.22).

It is now possible to evaluate the applicability of Perron's theorem [33], which asserts that for a system in the form

$$
\dot{\mathbf{y}}(t) = A\mathbf{y}(t) + \mathbf{g}(t, \mathbf{y}(t)), \quad (4.5.24)
$$

if $\sigma(A) \subset \mathbb{C}^-$ and

$$
\lim_{\|\mathbf{y}\| \to 0} \frac{\|\mathbf{g}(t, \mathbf{y})\|}{\|\mathbf{y}\|} = 0 \quad (4.5.25)
$$

uniformly with respect to t, then $\mathbf{y} = \mathbf{0}$ is exponentially asymptotically stable.

Given that

$$
\|\mathbf{g}(\mathbf{y})\| = \|C(\mathbf{y} \circ \mathbf{y})\|, \quad (4.5.26)
$$

condition (4.5.25) is satisfied. Consequently, if $\sigma(\tilde{A}) \subset \mathbb{C}^-$ also holds, the system exhibits exponential asymptotic stability.

Assuming A, C, and $\mathbf{b}$ are known for a specific instance of system (4.5.16), it becomes necessary to compute an approximation of the equilibrium point $\bar{\mathbf{y}}$ and, subsequently, $\tilde{A}$:

$$
\tilde{A} = A + 2CD_{\bar{\mathbf{y}}}, \quad D_{\bar{\mathbf{y}}} = \operatorname{diag}(\bar{\mathbf{y}}). \quad (4.5.27)
$$

Only after this approximation can the stability condition be verified. In the specific case of system (4.5.9)–(4.5.14), where C has only one nonzero element, the expression simplifies to:

$$2CD_{\bar{y}} = 2(c_{46}\mathbf{e}_4)(\bar{y}_6\mathbf{e}_6^T) = 2\beta_{SNcVTA}^{LC}\bar{y}_6\mathbf{e}_4\mathbf{e}_6^T, \tag{4.5.28}$$

where $\mathbf{e}_k$ denotes the k-th versor of the canonical basis.

The equilibrium point $\bar{\mathbf{y}}$ can be approximated using the Newton method to find the root of:

$$f(\mathbf{y}) = A\mathbf{y} + C(\mathbf{y} \circ \mathbf{y}) + \mathbf{b}. \tag{4.5.29}$$

The iteration follows:

$$\begin{aligned}\bar{\mathbf{y}}^{l+1} &= \bar{\mathbf{y}}^l - f'(\bar{\mathbf{y}}^l)^{-1}f(\bar{\mathbf{y}}^l) \\ &= \bar{\mathbf{y}}^l - (A + 2CD_{\bar{y}^l})^{-1}f(\bar{\mathbf{y}}^l).\end{aligned} \tag{4.5.30}$$

The initial point $\bar{\mathbf{y}}^0$ serves as the solution to $A\bar{\mathbf{y}} + \mathbf{b} = \mathbf{0}$. The system is exponentially asymptotically stable if both conditions $\sigma(A) \subset \mathbb{C}^-$ and $\sigma(\tilde{A}) \subset \mathbb{C}^-$ are satisfied.

The iteration to compute the equilibrium point can terminate when:

$$\max_i |\bar{y}_i^{l+1} - \bar{y}_i^l| < \text{tol}, \tag{4.5.31}$$

indicating that the maximum error across each component of $\bar{\mathbf{y}}^{l+1}$ falls below a specified tolerance.

4.5.3.2   Reference Data

The average brain activation in healthy subjects, as well as the time constants, have been extracted from literature. Starting from these average values, we craft a synthetic activity profile for a population of subjects (240 simulated subjects). This is achieved by modeling a normal distribution centered around the average value, with a normalized maximum excursion of $\pm 50\%$. In the reference (or target) data that we aim to replicate using the computational model, a population of adult male rats is subdivided into six groups:

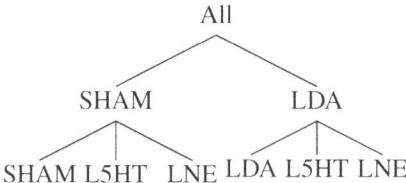

where:

- SHAM: indicates that subjects are treated with saline;
- LDA: dopamine depletion;
- LNE: noradrenaline depletion;
- L5HT: serotonin depletion.

Since there is no single study that lists all the required brain areas activation values for a particular subject at the same time, we have no choice but to generate a synthetic population of virtual subjects with area activation values which lie within the distributions identified across the literature. The generated target values distribution for all cases are summarized in Table 4.1 and Fig. 4.5.

Table 4.1 Target values

Area	Value
GP	$\leftarrow \mathcal{N}(22, 22 \cdot \frac{1}{8})$
GP^{LDA}	$= GP$
GP^{L5HT}	$= GP \cdot 0.65$
GP^{LNE}	$= GP$
$GP^{LDA+L5HT-max}$	$= GP \cdot 0.75$
$GP^{LDA+L5HT-min}$	$= GP \cdot 0.65$
$GP^{LDA+LNE-max}$	$= GP$
$GP^{LDA+LNE-min}$	$= GP \cdot 0.65$
StrD1	$\leftarrow \mathcal{N}(10, 10 \cdot \frac{1}{8})$
StrD2	$\leftarrow \mathcal{N}(9, 9 \cdot \frac{1}{8})$
SNcVTA	$\leftarrow \mathcal{N}(4.47, 4.47 \cdot \frac{1}{8})$
$SNcVTA^{LDA}$	$= SNcVTA \cdot 0.1$
DRN	$\leftarrow \mathcal{N}(1.41, 1.41 \cdot \frac{1}{8})$
DRN^{L5HT}	$= DRN \cdot 0.3$
LC	$\leftarrow \mathcal{N}(2.3, 2.3 \cdot \frac{1}{8})$
LC^{LDA}	$= LC \cdot 0.8$
LC^{LNE}	$= LC \cdot 0.2$

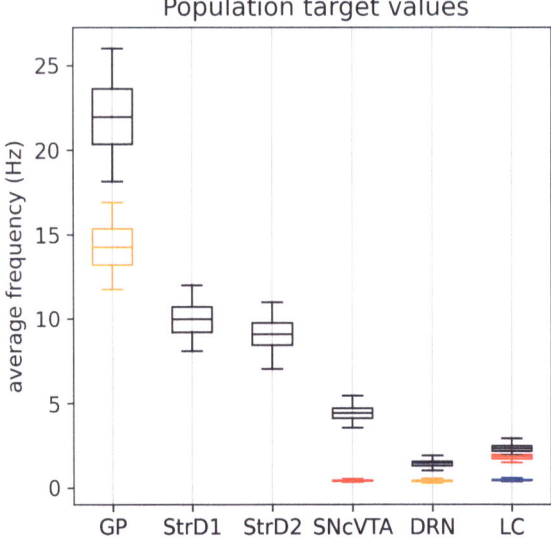

Fig. 4.5 Generated target values distribution for all cases: in black, the SHAM case. The LDA dopaminergic lesion target (red) differs from SHAM only for SNcVTA and LC. The serotonergic L5HT lesion target (yellow) differs from SHAM only in for GP and DRN. Finally, the noradrenergic LNE lesion target (blue) differs from SHAM only for LC. Values for each individual are generated so that each area follows a normal distribution around a center value with a maximum spread of $\pm 50\%$ ($4\sigma = 0.5\mu$). Adapted with permission from [3]

4.5.3.3   Model Fitness

The model will be used to make predictions on the behaviour of combined depletions; we therefore require the model to be able to reproduce its target data in four different states at the same time: healthy (SHAM), dopaminergic lesion (LDA), noradrenergic lesion (LNE) and serotonergic lesion (L5HT). The combinations LDA+LNE and LDA+L5HT are instead constrained only to a target range, to be able to also serve as a prediction (and hence as a measure of the agreement of the model with experimental data). Let S_i be the set of parameters that define the model representing test subject i. S_i contains:

- 6 time constants: τ_{GP}, τ_{StrD1}, τ_{StrD2}, τ_{SNcVTA}, τ_{DRN}, τ_{LC}. time constants are derived from literature and are therefore not optimized;
- SHAM: the healthy model has 20 free parameters, namely all α and β parameters defined in system (4.5.9)–(4.5.14);
- LDA: the dopaminergic lesion instance has 4 free parameters, i.e. α_{SNcVTA}^{DRN}, α_{SNcVTA}^{LC}, β_{SNcVTA}^{LC}, α_{SNcVTA}^{ext}. Those are all the parameters of the SNcVTA equation. All the other parameters are kept constant and are the same as in SHAM;
- L5HT: the serotonergic lesion instance has 3 free parameters, i.e. α_{DRN}^{SNcVTA}, α_{DRN}^{LC}, α_{DRN}^{ext}, All the other parameters are kept constant and are the same as in SHAM;

- LNE: the Noradrenergic lesion has 3 free parameters. i.e. α_{LC}^{SNcVTA}, α_{LC}^{DRN}, α_{LC}^{ext}, All the other parameters are kept constant and are the same as in SHAM;
- LDA+L5HT, LDA+LNE: the combination of lesions do not have any free parameters but constructed by applying to the SHAM values, in order, the relevant values from each lesion.

The set S_i therefore contains a total of 36 parameters, 30 of which must be optimized at the same time to fit the available data. Appropriate subsets of the parameters in S_i are then used to build the corresponding matrices A, C, $\mathbf{b}$ to completely define system (4.5.16), and hence compute its solution and properties.

The set S_i therefore contains a total of 36 parameters, 30 of which must be optimized at the same time to fit the available data. Appropriate subsets of the parameters in S_i are then used to build the corresponding matrices A, C, $\mathbf{b}$ to completely define system (4.5.16), and hence compute its solution and properties.

We will hereafter refer to S_i as the complete model for subject i, since it is the set of parameters that completely define it. The variations S_i^{kind}, like S_i^{SHAM}, S_i^{LDA} and so on, will refer instead to the subset of parameters which are currently being applied to actually simulate the model.

We will denote one solution as:

$$S_i^{SHAM}(\mathbf{y}_0, t_0, T) = Y = \begin{pmatrix} y_1(t_0) & \cdots & y_1(t_N) \\ \vdots & & \vdots \\ y_s(t_0) & \cdots & y_s(t_N) \end{pmatrix} \qquad (4.5.32)$$

Y is therefore the solution obtained by integrating the model in the interval $[t_0, T]$, with the starting vector $\mathbf{y}_0$, and using the SHAM subset of parameters. The matrix Y comprises s rows, the i-th row corresponds to the i-th equation in the system of differential equations.

The number N of integration steps, as well as their size, is usually variable and chosen by the integration method case-by-case, hence it can potentially be different for each subset of parameters.

Likewise, we will denote with T_i^{kind} the corresponding reference solutions that will be used to evaluate the fitness of the model:

$$T_i^{SHAM}(J) = Y_T(J) = \begin{pmatrix} y_{T1}(t_0) & \cdots & y_{T1}(t_N) \\ \vdots & & \vdots \\ y_{Ts}(t_0) & \cdots & y_{Ts}(t_N) \end{pmatrix} \qquad (4.5.33)$$

where $J = [t_0, ..., t_n]$ is a vector of times. The fitness of a model is finally obtained as a combination of many fitness figures $f_i \in [0, 1]$, which measure a wide range of properties of the simulated solution with respect to the reference ones.

More details about the fitness The fitness measure for subject S_i is therefore a composition of many fitness. To completely describe the model of a subject, other than its set of parameters S_i, we need to represent its corresponding set of target values

T_i: T_i contains the reference solutions that the model is supposed to reproduce when using the parameters in S_i; T_i must therefore have one reference solution for each of the states (healthy and lesions) that we are modelling.

In particular, we assume that we can address a particular reference solution from T_i in a similar way to the solution corresponding to particular subsets of parameters in S_i, as described in (4.5.33), under the assumption that T_i can provide the reference solution for any discrete set of times $J = \{t_0, ..., t_N\}$ which is decided by the integration algorithm during the computation of the solution $S_i^{\text{SHAM}}(\mathbf{y}_0, t_0, T)$. The same notation of course applies for the other cases, T_i^{LDA}, T_i^{L5HT} and so on.

The subject index will intentionally be left out in the following sections to lighten the notation further, since it's not relevant in the context: the fitness is of course computed independently for each subject in the same way.

SHAM fitness The fitness of the healthy instance is divided in one fitness measure for each equation.

To simplify the notation, we define:

- $Y_T = T^{SHAM}(J)$, the corresponding target solution
- $\mathbf{y}_0 = T^{SHAM}(t_0)$ the starting vector
- $Y = S^{SHAM}(\mathbf{y}_0, t_0, T)$, the simulation of the model using the appropriate subset of parameters
- $J = \{t_0, ..., t_N \leq T\}$, the time base chosen by the integration method.

For each of the s equations in the status vector we can compute the corresponding mse_i:

$$\text{mse}_i = \sum_{j=1}^{N}(t_j - t_{j-1})e_{ij}^2, \quad (e)_{ij} = Y_T - Y, \quad i = 1, .., s \qquad (4.5.34)$$

and finally the simulation-time-weighted fitness:

$$f_i^{SHAM} = \frac{t_N - t_0}{T - t_0} \frac{1}{1 + mse_i} \qquad (4.5.35)$$

The set of measures for the SHAM instance is therefore:

$$F^{SHAM} = \left\{ f_i^{SHAM} | i = 1, ..., s \right\} \qquad (4.5.36)$$

LDA fitness Similarly to the SHAM case, we define:

- $Y_T = T^{LDA}(J)$, the corresponding target solution
- $\mathbf{y}_{h0} = T^{SHAM}(t_0)$ the starting vector for the healthy case
- $\mathbf{y}_{l0} = T^{LDA}(t_0)$ the starting vector for the LDA case
- $Y = S^{LDA}(\mathbf{y}_0, t_0, T)$, the simulation of the model using the appropriate subset of parameters
- $J = \{t_0, ..., t_N \leq T\}$, the time base chosen by the integration method
- $t_c = \frac{T - t_0}{2}$, the time before which we ignore the solution's fitness.

In this case we have only three reference solutions to consider: GP,SNcVTA and LC; in particular, the former is to be fitted exactly from experimental data. The SNcVTA fitness is a conceptual requirement since we don't have the corresponding exact experimental data: LDA is a lesion of neurons in SNcVTA that in turn lowers the levels of dopamine. We therefore require that the average activation frequency of SNcVTA has to become at least as low as indicated in the reference solution, but can be free to become even lower. Similarly, available data suggests a lowered activity in LC to be at most 80% of the healthy value.

We also require the solution to be stable with two different initial conditions: the solution should obviously be stable near the equilibrium point (the LDA solution $\mathbf{y}_{l0}$), but perhaps more importantly, a healthy subject (hence starting with $\mathbf{y}_{h0}$) must be able to transition to the lesioned state (as it naturally occurs in-vivo during the experiment) without incurring in instabilities. This also implies that the first transient phase should be ignored in the computation of the fitness; for that reason we defined t_c as a threshold time before which we ignore the solution. Particular care should be used in choosing t_0, T and consequently t_c big enough with respect to the time constants of the system.

We therefore define two fitness measures for GP:

$$f_{GP}^{LDA}(\mathbf{y}_{l0}, t0, T, t_c), \quad f_{GP}^{LDA}(\mathbf{y}_{h0}, t0, T, t_c) \tag{4.5.37}$$

both are computed as in (4.5.36):

$$f_{GP}^{LDA}(\mathbf{y}_{l0}, t0, T, t_c), \quad f_{GP}^{LDA}(\mathbf{y}_{h0}, t0, T, t_c) \tag{4.5.38}$$

$$f_i^{LDA}(\mathbf{y}, t_0, T, t_c) = \frac{t_N - t_0}{T - t_0} \frac{1}{1 + mse_i}, \tag{4.5.39}$$

$$mse_i = \sum_{j=c}^{N} (t_j - t_{j-1}) e_{ij}^2, \quad (e)_{ij} = Y_T - Y, \tag{4.5.40}$$

where Y, Y_T and hence J are of course computed accordingly to the selected $\mathbf{y}$, c is the index of the first $t \geq t_c$ in J, and the index of GP in the status vector happens to be 1, hence $i = 1$.

The fitness f_{SNcVTA}^{LDA} for SNcVTA is computed in a similar way, but also applying a sieve function. that ignores negative errors:

$$f_i^{LDA}(\mathbf{y}, t_0, T, t_c) = \frac{t_N - t_0}{T - t_0} \frac{1}{1 + mse_i}, \tag{4.5.41}$$

$$mse_i = \sum_{j=c}^{N} (t_j - t_{j-1}) \max(0, e_{ij})^2, \quad (e)_{ij} = Y_T - Y, \tag{4.5.42}$$

where i is the index of the SNcVTA equation in the status vector.

Following the same principles, we define the two fitness measures for LC:

$$f_i^{LDA}(\mathbf{y}, t_0, T, t_c) = \frac{t_N - t_0}{T - t_0} \frac{1}{1 + mse_i}, \tag{4.5.43}$$

$$mse_i = \sum_{j=c}^{N} (t_j - t_{j-1}) \max(0, e_{ij})^2, \quad (e)_{ij} = Y_T - Y, \tag{4.5.44}$$

where this time i is the index of the LC equation in the status vector.

The set of measures for the LDA case therefore has four elements:

$$\begin{aligned}
F^{LDA} = \big\{ & f_{GP}^{LDA}(\mathbf{y}_{l0}, t0, T, t_c), f_{GP}^{LDA}(\mathbf{y}_{h0}, t0, T, t_c), \\
& f_{SNcVTA}^{LDA}(\mathbf{y}_{l0}, t0, T, t_c), f_{SNcVTA}^{LDA}(\mathbf{y}_{h0}, t0, T, t_c), \\
& f_{LC}^{LDA}(\mathbf{y}_{l0}, t0, T, t_c), f_{LC}^{LDA}(\mathbf{y}_{h0}, t0, T, t_c) \big\}
\end{aligned} \tag{4.5.45}$$

L5HT and LNE fitness The fitness measures for this two instances are conceptually identical to the LDA case; the fitness for GP is therefore computed according to (4.5.39), and the fitness for the lesioned areas use the the same sieve as in (4.5.41) (but of course selecting the correct lesioned area, respectively DRN and LC). The sets of measures for this two instances are defined as:

$$\begin{aligned}
F^{L5HT} = \big\{ & f_{GP}^{L5HT}(\mathbf{y}_{l0}, t0, T, t_c), f_{GP}^{L5HT}(\mathbf{y}_{h0}, t0, T, t_c), \\
& f_{DRN}^{L5HT}(\mathbf{y}_{l0}, t0, T, t_c), f_{DRN}^{L5HT}(\mathbf{y}_{h0}, t0, T, t_c) \big\}
\end{aligned} \tag{4.5.46}$$

$$\begin{aligned}
F^{LNE} = \big\{ & f_{GP}^{LNE}(\mathbf{y}_{l0}, t0, T, t_c), f_{GP}^{LNE}(\mathbf{y}_{h0}, t0, T, t_c), \\
& f_{LC}^{LNE}(\mathbf{y}_{l0}, t0, T, t_c), f_{LC}^{LNE}(\mathbf{y}_{h0}, t0, T, t_c) \big\}
\end{aligned} \tag{4.5.47}$$

Lesion combination fitness The combination of lesions is left as unconstrained as possible to be able to serve as a prediction; we do however at least require the corresponding simulation not to diverge and to lie within an acceptable range. In the experiment lesions are applied in succession, LDA always first. It makes sense to require the solution to be stable with both $\mathbf{y}_{h0} = T^{SHAM}(t_0)$ and $\mathbf{y}_{l0} = T^{LDA}(t_0)$ as initial conditions. We define a fitness which only considers the temporal span of the solution to penalize early divergence:

$$f^{LDA+LNE}(\mathbf{y}, t_0, T) = \frac{t_N - t_0}{T - t_0}, \quad f^{LDA+L5HT}(\mathbf{y}, t_0, T) = \frac{t_N - t_0}{T - t_0} \tag{4.5.48}$$

where J and consequently t_N, t_0 are computed from the corresponding simulation $S^{LDA+lesion}(\mathbf{y}, t_0, T)$.

Additionally, we require the GP value of LDA+L5HT to be within reasonable limits. In particular, we define specific limit fitness functions similar to (4.5.41), for example the lower bound:

$$f_{i,\text{min}}^{LDA+L5HT}(\mathbf{y}, t_0, T, t_c) = \frac{t_N - t_0}{T - t_0} \frac{1}{1 + mse_i}, \tag{4.5.49}$$

$$mse_i = \sum_{j=c}^{N} (t_j - t_{j-1}) \min(0, e_{ij})^2, \quad (e)_{ij} = Y_T - Y, \tag{4.5.50}$$

$$Y_T = T^{LDA+L5HT-\text{min}}(J) \tag{4.5.51}$$

where i is again the index corresponding to GP and $T^{LDA+L5HT-\text{min}}(J)$ is the reference solution containing the lower bound; likewise the upper bound will be defined similarly but using max as an error sieve against the upper bound reference solution $T^{LDA+L5HT-\text{max}}(J)$. This limits are necessary to guide the optimization towards a solution which lies within the experimentally determined range and exclude instead solutions which may exhibit better fitness scores but are outside of the physiological range.

The set of measures for the combination of lesions is therefore:

$$\begin{aligned} F^{COMB} = \big\{ &f^{LDA+LNE}(\mathbf{y}_{h0}, t_0, T), f^{LDA+LNE}(\mathbf{y}_{l0}, t_0, T), \\ &f^{LDA+L5HT}(\mathbf{y}_{h0}, t_0, T), f^{LDA+L5HT}(\mathbf{y}_{l0}, t_0, T), \\ &f_{GP,\text{min}}^{LDA+LNE}(\mathbf{y}_{h0}, t_0, T), f_{GP,\text{min}}^{LDA+LNE}(\mathbf{y}_{l0}, t_0, T), \\ &f_{GP,\text{max}}^{LDA+LNE}(\mathbf{y}_{h0}, t_0, T), f_{GP,\text{min}}^{LDA+LNE}(\mathbf{y}_{l0}, t_0, T), \\ &f_{GP,\text{max}}^{LDA+L5HT}(\mathbf{y}_{h0}, t_0, T), f_{GP,\text{min}}^{LDA+L5HT}(\mathbf{y}_{l0}, t_0, T), \\ &f_{GP,\text{max}}^{LDA+L5HT}(\mathbf{y}_{h0}, t_0, T), f_{GP,\text{max}}^{LDA+L5HT}(\mathbf{y}_{l0}, t_0, T) \big\} \end{aligned} \tag{4.5.52}$$

Parameters constraints The fitness function can also be useful to impose soft, dynamic constraints on the parameters. In this case, it makes sense to require the α^{ext} parameters of a lesion to be less or equal than its counterpart in the SHAM instance: that particular brain area have been damaged, and it makes sense to assume it would lower its average activation frequency in absence of other stimuli.

We define the fitness measure:

$$f_l^{PAR} = \frac{1}{1 + max(0, S^l - S^{SHAM})} \tag{4.5.53}$$

where l is one of the three lesions (LDA, LNE, L5HT) and $S^l - S^{SHAM}$ represent the difference between the altered α^{ext} parameter in the lesioned subset and its counterpart in the healthy one.

As usual, we define the set:

$$F^{PAR} = \big\{ f_{LDA}^{PAR}, f_{LNE}^{PAR}, f_{L5HT}^{PAR} \big\} \tag{4.5.54}$$

Asymptotic stability constraints Every subject state considered in this study is supposed to be stable in time; it is therefore important to impose that each set of parameters defines an asymptotically stable system which will ultimately never diverge from its equilibrium point.

As discussed previously, system (4.5.9)–(4.5.14) is exponentially asymptotically stable if $\sigma(\tilde{A}) \subset \mathbb{C}^-$, where $A = A + 2CD_{\bar{y}}$ (see (4.5.27)).

We can therefore enforce a fitness measure:

$$f_l^{STAB} = \frac{1}{1 + \sum_i \max(0, \mathbb{R}(\lambda_i))} \tag{4.5.55}$$

where l is one of the parameter subsets which define the model (SHAM, LDA, etc.), and λ_i is an eigenvalue of the corresponding $\tilde{A}$. This measure will therefore always be 1 when the system is asymptotically stable, but tend to zero as the real part of the eigenvalues grows more positive.

The computation of $\tilde{A}$ requires using an iterative root-finding method to determine the equilibrium point $\bar{y}$ of each parameters set $A, C, \mathbf{b}$. It is advantageous to use multiple stopping conditions for this method to avoid unnecessary computation, in particular:

- A tolerance on the precision of $\bar{y}^l$ as defined in (4.5.31); this tolerance should be set to be compatible with the precision obtained with the parameters optimization algorithm. For example, if the optimization fitness required translates to an mse of 10^{-8}, it makes sense to require tol $= 10^{-9}$.
- A arbitrary guard on the maximum number of allowed iterations; since precision is not of paramount importance in this context, the number of iterations can be kept rather small (≤ 25).
- A guard on the value of the components of $\bar{y}$. If the method is converging to an equilibrium point which has some components which are too big or negative, the stability of the system is ultimately meaningless in the context of this study, therefore it is not worth investing computing power in obtaining it with high precision.

We finally define the set:

$$F^{STAB} = \{ f_{SHAM}^{STAB}, f_{LDA}^{STAB}, f_{LNE}^{STAB}, f_{L5HT}^{STAB}, \\ f_{LDA+LNE}^{STAB}, f_{LDA+L5HT}^{STAB} \} \tag{4.5.56}$$

The fitness, at last We now have all the elements needed to compute the fitness of subject S_i.

Let F be the union of all the sets of measures we have defined:

$$F = F^{SHAM} \cup F^{LDA} \cup F^{L5HT} \cup F^{LNE} \cup F^{COMB} \cup F^{PAR} \cup F^{STAB} \tag{4.5.57}$$

we can now combine all the fitness measures we identified to obtain the total (or final) fitness figure:

$$f = \sqrt{\min_i(f_i)\frac{1}{n}\sum_{i=1}^{n}f_i}, \quad f_i \in F, \ n = |F| \qquad (4.5.58)$$

4.5.3.4 Simulation and optimization

All simulations are obtained using a variable order, variable step scheme backward-differentiation formulas (BDF) [69] integrator provided by the Python scipy library. The optimization of parameters is then performed using Differential Evolution (DE), also as provided by the Python scipy package, with a population of 240 simulated subjects, DE/best/1/exp strategy with $C_r = F = 0.95$ initialized with a uniform Halton distribution.

An external optimization cycle which sets different random generator seeds have been used to retry the cases which did not find convergence. In so doing, all cases did eventually converge after a few attempts. The full codebase can be found at https://github.com/WohthaN/Simulating_noradrenaline_and_serotonin_depletions_in_parkinson.

4.5.3.5 Results

All models fit the corresponding target values as defined in Table 4.1 with a fitness $f \geq 1 - 10^{-8}$. Since the measure is dominated by the smallest fitness value being combined by definition, also the mean square difference of each component of the solution from its reference value is bonded by the same order of magnitude, which is a suitable precision for the purposes of this work. Figure 4.6 shows an overview of the simulated behaviour of all six areas in all conditions. In all cases, the simulated values for the SHAM case overlap the target values with the imposed tolerance. In the lesion groups, all areas which do not have a target defined, are model predictions. All instances of the model meet the stability conditionsas imposed by the fitness measure. Figure 4.7 shows an example of dynamic behaviour of one of the models.

4.5.3.6 Exploring Potential Treatment Avenues

The computational model offers insights into the key parameters that govern the activity of simulated brain regions. Researchers conducted a sensitivity analysis for each simulated brain area concerning the various model parameters. Each parameter can vary independently (within a defined range) to assess its effects on the simulated brain areas. Analysts can replicate similar observations for all individuals within the available population, allowing for a comparison of the average excursion of each

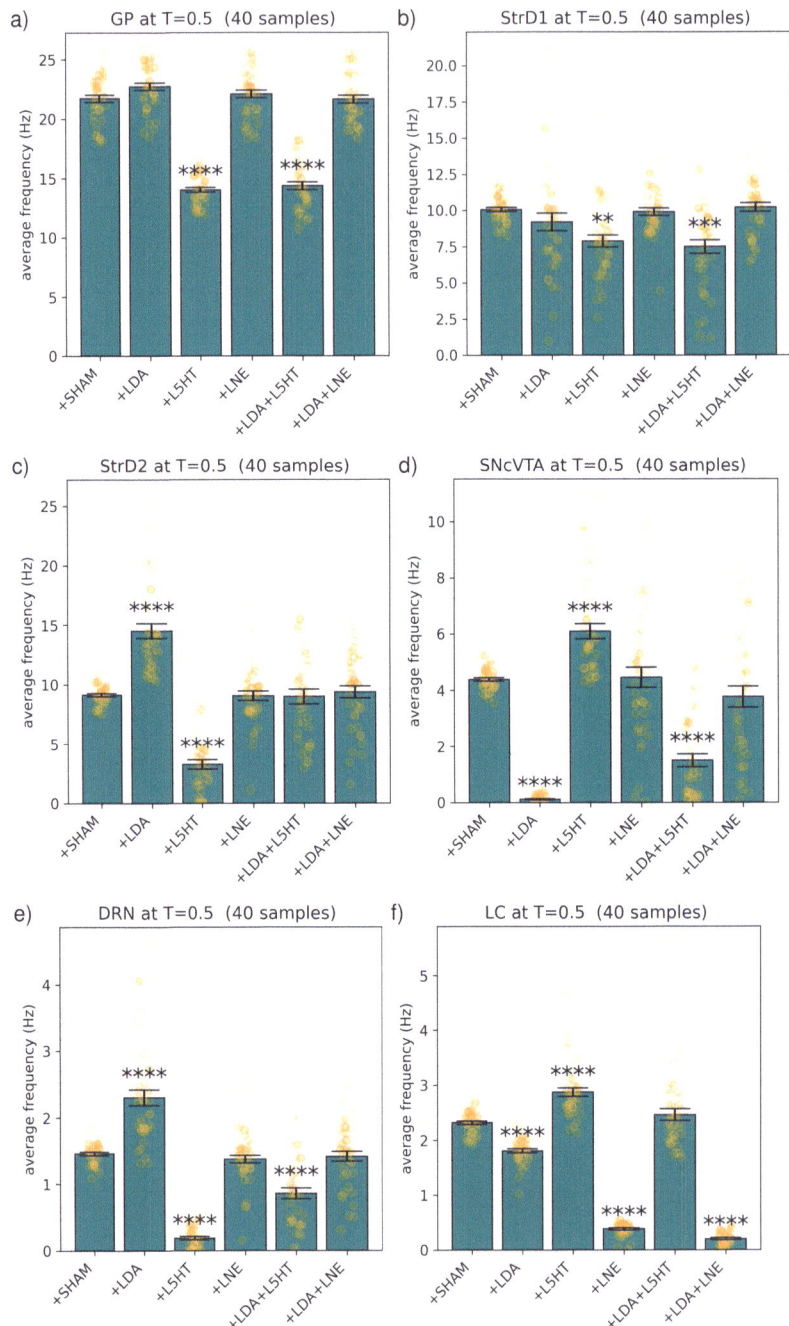

Fig. 4.6 Summary of the behavior exhibited by GP (**a**), StrD1 (**b**), StrD2 (**c**), SNcVTA (**d**), DRN (**e**), and LC (**f**) across various groups including SHAM, LDA, L5HT, LNE, LDA+L5HT, and LDA+LNE, with each group consisting of 40 distinct subjects binned accordingly. Adapted with permission from [3]

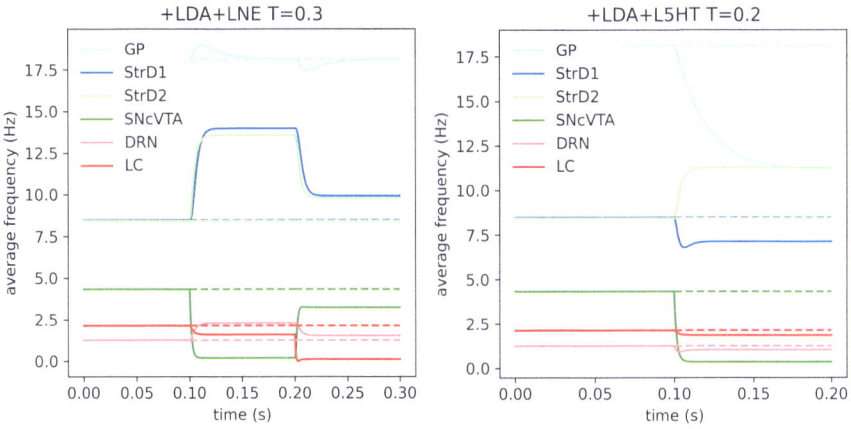

Fig. 4.7 Example of dynamic behaviour of a simulated target. On the left, lesions are applied one by one at different times, while on the right, lesions are applied simultaneously. The dynamic behaviour is different, but as the systems have been chosen by the fitness function to be asymptotically stable in all conditions, the same stable solution is reached in both cases

area with the average excursion of the parameter. This comparison helps infer the relative importance of each parameter. For each individual in the population and for each parameter, the following steps occur:

- Vary the parameter around its original value by $\pm 50\%$ in 100 uniform increments. If v represents the parameter value for individual S_i^{SHAM}, define the resulting set as:

$$V_{param,i} = \left\{ \left(\frac{x}{99} + \frac{97}{198} \right) v \right\} \subseteq [0.5v, 1.5v], \quad i = 1, \ldots, 100.$$

This generates a distinct $V_{param,i}$ set for each parameter of each individual.
- Simulate the model using each value in $V_{param,i}$, and record the final activation value for each brain area. Discard results from simulations that terminate prematurely-indicating divergence or the attainment of physically impossible states-and remove the corresponding parameter value from $V_{param,i}$. As a result, obtain six sets for each parameter and individual, one for each brain area, denoted as $A_{param,i}^{area}$.

Combine the sets across the population:

$$A_{param}^{area} = \bigcup_i A_{param,i}^{area}, \quad V_{param} = \bigcup_i V_{param,i}, \quad (4.5.59)$$

where $param \in S^{SHAM}$ represents one of the free parameters, $area$ corresponds to one of the six brain areas $\{GP, StrD1, StrD2, SNcVTA, DRN, LC\}$, and i refers to the i-th subject in the population.

A sensitivity index computes for each area by scaling both V and A with their respective median values and dividing the standard deviations:

$$I_{param,area} = \frac{\text{std}(A^{area}_{param}/\text{median}(A^{area}_{param}))}{\text{std}(V_{param}/\text{median}(V_{param}))} \qquad (4.5.60)$$

The sensitivity index $I_{param,area}$ represents a *sensitivity matrix*, where each column corresponds to a brain area and each row corresponds to a free parameter. Normalization of $I_{param,area}$ occurs with respect to its maximum value in the final step.

Figure 4.8 displays the computed sensitivity matrix for the entire fitted population in the SHAM case. A value of 1 signifies the maximum measured sensitivity, while a value of 0 indicates that a specific parameter exerts no effect on that area.

The matrix clearly shows that all areas exhibit relative sensitivity to changes in noradrenergic balance (external activation of the locus coeruleus, LC). Additionally, a somewhat lesser sensitivity appears for changes in serotonergic balance (external activation of the dorsal raphe nucleus, DRN). Consequently, stimulating the LC and/or DRN likely induces changes in the activation of all brain areas.

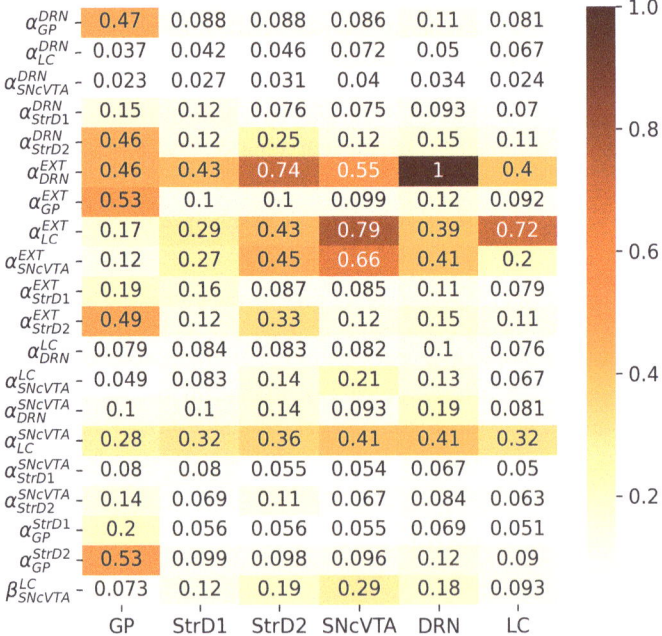

Fig. 4.8 The comparative sensitivity of each brain area to individual parameters among healthy subjects is analyzed. A higher value indicates a greater impact (in absolute magnitude) of a parameter on the activation frequency of a given area. The values are normalized to unity, emphasizing the relative magnitude of effects derived from each parameter. Adapted with permission from [3]

According to the model depicted in Fig. 4.4, dopaminergic levels can potentially alter in two ways:

1. External stimulation of the locus coeruleus (LC) modifies its production of noradrenaline.
2. External stimulation of the dorsal raphe nucleus (DRN) modifies its production of serotonin.

Stimulation can occur either chemically, by supplying the necessary precursors for monoamine synthesis, or electrically, to artificially adjust the average firing rate of neurons in the targeted area, thereby increasing monoamine production and projecting more of these neurotransmitters to the connected brain regions.

The sensitivity matrix in Fig. 4.8 indicates that while LC stimulation likely has a significant impact on dopamine levels, DRN stimulation probably has a lesser effect on the activation of the substantia nigra pars compacta and ventral tegmental area complex (SNcVTA). This outcome may lead to adverse side effects that could compromise effective treatment.

Researchers tested these hypotheses by optimizing stimulation parameters applied to the targeted areas, DRN and LC. The parameter sets for each subject expanded to include three new fitness measures:

- LDA + cLC with free parameters α_{LC}^{ext} and corresponding fitness measure F^{cLC}.
- LDA + cDRN: α_{DRV}^{ext} with fitness measure F^{cDRN}.
- LDA + cCOMB: α_{LC}^{ext} and α_{DRV}^{ext} with fitness measure F^{cCOMB}.

For each subject, the corresponding model matrices A, C, and $\mathbf{b}$ derive from the LDA (hence dopamine-depleted) parameter set, leaving the external stimulation of the tested areas as the only free parameters. The different fitness measures for the three populations ensure that each one stimulates a unique area to simulate treatment. The stimulated area does not contribute to the respective fitness measure, which focuses on:

- The mean squared error of the activation values for each area, one measure per area, as defined for the SHAM case in Eq. (4.5.35), excluding the stimulated area (for instance, excluding LC).
- A parameter constraint, similar to that defined in Eq. (4.5.53), which enforces the external stimulus parameter to equal or exceed the original value. This constraint encourages optimization to favor stimulation rather than inhibition. Specifically, the component defines as:

$$f_{cLC}^{PAR} = \frac{1}{1 + \max(0, S^{SHAM} - S^{cLC})}. \tag{4.5.61}$$

- An asymptotic stability constraint as defined in Eq. (4.5.55), where $\tilde{A}$ constructs using the current subset of parameters $S^{LDA+cLC}$.

The fitness measures F^{cDRN} and F^{cCOMB} follow an analogous construction, with mean squared errors for both stimulated areas omitted from the measure.

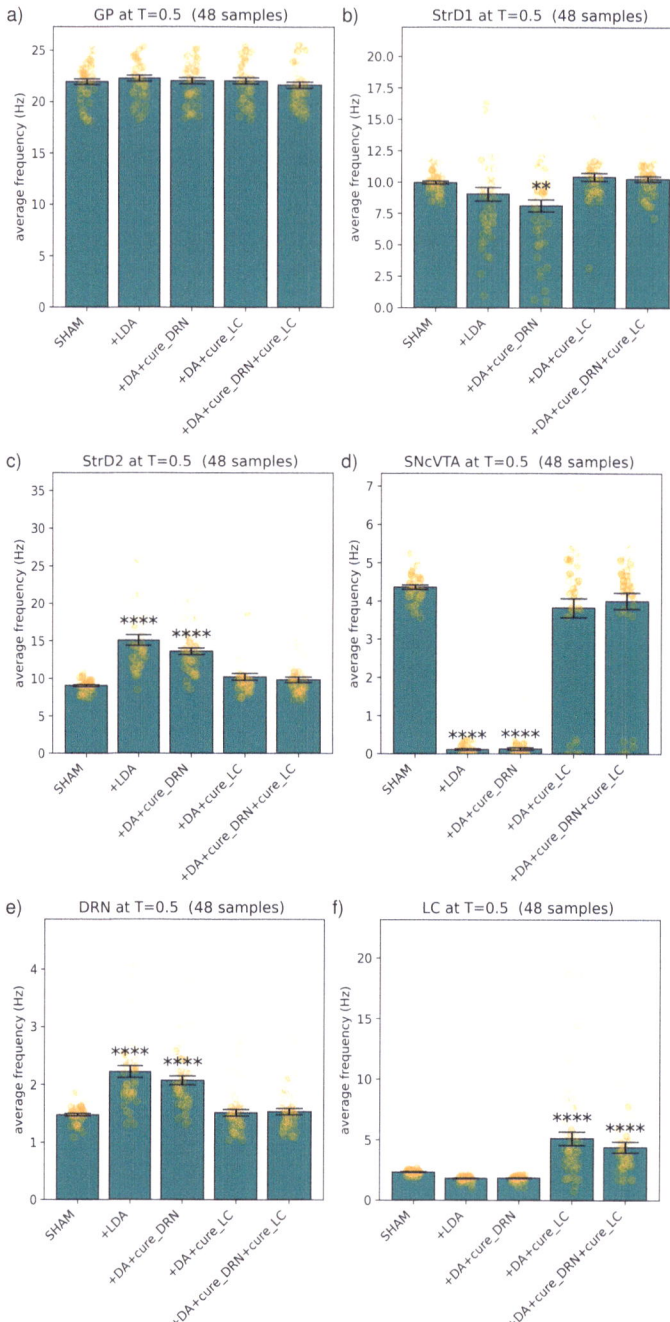

Fig. 4.9 Lesion and treated values for GP (**a**), StrD1 (**b**), StrD2 (**c**), SNcVTA (**d**), DRN (**e**), LC (**f**). A statistically significant increase in average LC activity (and thus noradrenaline levels) can restore the activity (and consequently monoamine production levels) of all areas significantly affected by LDA to SHAM levels. Adapted with permission from [3]

Optimization successfully restores healthy activation levels in most subjects by stimulating either the LC or both the LC and DRN, but consistent failure occurs when only stimulating the DRN. Figure 4.9 provides compelling evidence of the significant impact of statistically meaningful LC stimulation on re-establishing equilibrium between serotonin and dopamine levels, as reflected in the activation levels of the DRN and SNcVTA, respectively, within the population of LDA subjects.

Conclusions

This book offers a comprehensive framework for constructing computational brain models through a four-step method that utilizes ordinary differential equations and Python programming. Recognizing the interdisciplinary nature of the audience, the authors have consciously maintained two levels of depth when covering mathematical concepts. This dual approach ensures accessibility for newcomers to the subject and those with more advanced STEM knowledge, balancing foundational explanations with more complex discussions. By doing so, the book aims to make complex topics like brain simulation approachable and engaging for all readers, regardless of their starting point. The computational models presented here serve as examples of valuable tools for operationalizing hypotheses about brain functions and dysfunctions, fostering innovative research approaches that traditional methods may not easily accommodate. Readers acquire the skills to design, simulate, visualize, and interpret brain models, enhancing understanding of the neural mechanisms underlying behavior and informing potential therapeutic strategies for neurological disorders. Ultimately, this book encourages exploration and collaboration across disciplines, aiming to broaden the scope of brain research and facilitate insights into the complex interplay between brain regions and their role in shaping behavior.

© The Editor(s) (if applicable) and The Author(s), under exclusive license to Springer   189
Nature Singapore Pte Ltd. 2025
D. Caligiore and S. Carli, *Simulating the Brain*, Brain Informatics and Health,
https://doi.org/10.1007/978-981-96-2718-9

Bibliography

1. Price KV, Storn RM, Lampinen JA (2005) Differential evolution: a practical approach to global optimization. Natural computing series, Springer, New York, Berlin
2. Caligiore D (2022) IA istruzioni per l'uso. il Mulino, Bologna
3. Carli S, Brugnano L, Caligiore D (2024) Simulating combined monoaminergic depletions in a PD animal model through a bio-constrained differential equations system. Front Comput Neurosci 18:1386841
4. Gamma E (ed) (1995) Design patterns: elements of reusable object-oriented software. Addison-Wesley professional computing series, Addison-Wesley, Reading, Mass
5. Percival H, Gregory B (2020) Architecture patterns with Python: enabling test-driven development, domain-driven design, and event-driven microservices, 1st edn. O'Reilly, Beijing, China, Boston, MA
6. Badenhorst W (2017) Practical python design patterns: pythonic solutions to common problems. Apress, New York
7. Humble J, Farley D (2010) Continuous delivery reliable software releases through build, test, and deployment automation. Pearson Education, Limited, Hoboken. OCLC: 1348485094
8. Martin RC, Feathers MC (2009) Clean code: a handbook of agile software craftsmanship. Prentice Hall, Upper Saddle River, NJ. OCLC: 297575371
9. Fowler M (2019) Refactoring: improving the design of existing code. Addison-Wesley signature series, 2nd edn. Addison-Wesley, Boston. OCLC: on1064139838
10. Freeman S, Pryce N (2010) Growing object-oriented software, guided by tests. The Addison-Wesley signature series, Addison-Wesley, Upper Saddle River, NJ, Munich
11. Winters T, Manshreck T, Wright H (2020) Software engineering at Google: lessons learned from programming over time, 1st edn. O'Reilly, Beijing, Boston, Farnham, Sebastopol, Tokyo
12. Spolsky J (2004) Joel on software: and on diverse and occasionally related matters that will prove of interest to software developers, designers, and managers, and to those who, whether by good fortune or ill luck, work with them in some capacity. Apress, Berkeley, CA
13. Spolsky J (2008) More Joel on software: further thoughts on diverse and occasionally related matters that will prove of interest to software developers, designers, and managers, and to those who, whether by good fortune or ill luck, work with them in some capacity. Apress, Berkeley, CA, New York, NY. Distributed to the book trade worldwide by Springer-Verlag. OCLC: ocn209723635
14. Cooper A (2004) The inmates are running the asylum. Sams, Indianapolis, IN

D. Caligiore and S. Carli, *Simulating the Brain*, Brain Informatics and Health, https://doi.org/10.1007/978-981-96-2718-9

15. Öggl B, Kofler M (2023) Git: project management for developers and DevOps teams, 1st edn. Rheinwerk Publishing, Boston

16. Kaufmann M, Dohmke T, Brown D (2022) Accelerate DevOps with GitHub: enhance software delivery performance with GitHub issues, projects, actions, and advanced security. Packt Publishing, Limited, Birmingham. OCLC: 1341442459

17. Narbski J (2016) Mastering Git: attain expert-level proficiency with Git for enhanced productivity and efficient collaboration by mastering advanced distributed version control features. Packt Publishing, Birmingham, UK. OCLC: 948621884

18. Eckel B (2012) Thinking in Java: the definitive introduction to object oriented programming in the language of the World Wide Web, 4th edn., 11th print edn. Prentice-Hall, Upper Saddle River, NJ

19. IEEE standard for floating-point arithmetic (2019) IEEE Std 754-2019 (Revision of IEEE 754-2008), pp 1–84

20. Wikipedia contributors (2023) Half-precision floating-point format—Wikipedia, the free encyclopedia. https://en.wikipedia.org/w/index.php?title=Half-precision_floating-point_format&oldid=1185884437

21. Epperson JF An introduction to numerical methods and analysis, Revised edn. Wiley-Interscience. OCLC: ocm85851582

22. Numerical recipes: the art of scientific computing. OCLC: ocn123285342

23. Gilat A, Subramaniam V Numerical methods for engineers and scientists: an introduction with applications using MATLAB. Wiley

24. Brugnano L, Magherini C, Sestini A (2019) Calcolo Numerico

25. Reusser R (2021) Half-precision floating-point, visualized. https://observablehq.com/@rreusser/half-precision-floating-point-visualized

26. https://evanw.github.io/float-toy/

27. https://numpy.org/doc/stable/index.html

28. Riley KF, Hobson MP, Bence SJ (2006) Mathematical methods for physics and engineering, 3rd edn. Cambridge University Press, Cambridge, New York. OCLC: ocm62532900

29. Galván ML (2017) The multivariate bisection algorithm

30. Bachrathy D, Stépán G (2012) Bisection method in higher dimensions, the efficiency number. Period Polytech Mech Eng 56(2):81–86

31. https://docs.scipy.org/doc/scipy/index.html

32. Lakshmikantham V, Trigiante D (2002) Theory of difference equations: numerical methods and applications. No. 251 in Monographs and textbooks in pure and applied mathematics, 2nd edn. Marcel Dekker, New York

33. Brugnano L (2021) Modelli numerici per la simulazione

34. Butcher JC (2016) Numerical methods for ordinary differential equations, 3rd edn. Wiley Blackwell, Chichester, West Sussex

35. Kelley WG, Peterson AC (2001) Difference equations: an introduction with applications, 2nd edn. Harcourt/Academic Press, San Diego

36. Gohberg I, Lancaster P, Rodman L (2009) Matrix polynomials. No. 58 in Classics in applied mathematics. Society for industrial and applied mathematics. Philadelphia, siam edn., [classics edn.]. OCLC: ocn311310303

37. Meyer CD (2000) Matrix analysis and applied linear algebra. Society for industrial and applied mathematics, Philadelphia

38. Nocedal J, Wright SJ (2006) Numerical optimization. Springer series in operations research, 2nd edn. Springer, New York. OCLC: ocm68629100

39. Luenberger DG, Ye Y (2008) Linear and nonlinear programming. No. 116 in International series in operations research and management science, 3rd edn. Springer, New York, NY

40. The SciPy community, scipy.optimize.differential_evolution. https://docs.scipy.org/doc/scipy-1.8.1/reference/generated/scipy.optimize.differential_evolution.html

41. Jeyakumar G, Shanmugavelayutham C (2011) Convergence analysis of differential evolution variants on unconstrained global optimization functions. Int J Artif Intell & Appl 2:116–127

42. Srb R, Knobloch R, Mlýnek J (2017) The classic differential evolution algorithm and its convergence properties. Appl Math 62(2):197–208
43. Bolie VW (1961) Coefficients of normal blood glucose regulation. J Appl Physiol 16:783–788
44. Ma M, Li J (2022) Dynamics of a glucose-insulin model. J Biol Dyn 16:733–745
45. Vejrazkova D, Vankova M, Lukasova P, Hill M, Vcelak J, Tura A, Chocholova D, Bendlova B (2023) The glycemic curve during the oral glucose tolerance test: is it only indicative of glycoregulation? Biomedicines 11:1278
46. Fröhlich F (2016) Network neuroscience. Academic Press
47. Caligiore D, Silvetti M, D'Amelio M, Puglisi-Allegra S, Baldassarre G (2020) Computational modeling of catecholamines dysfunction in Alzheimer's disease at pre-plaque stage. J Alzheimer's Dis 77(1):275–290
48. Caligiore D, Giocondo F, Silvetti M (2022) The Neurodegenerative Elderly Syndrome (NES) hypothesis: Alzheimer and Parkinson are two faces of the same disease. IBRO Neurosci Rep 13:330–343
49. Angelini G, Malvaso A, Schirripa A, Campione F, D'Addario SL, Toschi N, Caligiore D (2024) Unraveling sex differences in Parkinson's disease through explainable machine learning. J Neurol Sci 123091
50. Caligiore D, Helmich RC, Hallett M, Moustafa AA, Timmermann L, Toni I, Baldassarre G (2016) Parkinson's disease as a system-level disorder. Npj Parkinson's Dis 2:16025
51. D'Angelo E, Jirsa V (2022) The quest for multiscale brain modeling. Trends Neurosci 45(10):777–790
52. Montgomery RM (2023) Integrating scales in neuroscience: a comprehensive review of microscale, mesoscale, and macroscale brain dynamics
53. Liu M, Fang S, Dong H, Xu C (2021) Review of digital twin about concepts, technologies, and industrial applications. J Manuf Syst 58:346–361
54. Semeraro C, Lezoche M, Panetto H, Dassisti M (2021) Digital twin paradigm: a systematic literature review. Comput Ind 130:103469
55. Caligiore D (2024) Curarsi con l'intelligenza artificiale. il Mulino, Bologna
56. Schwartenbeck P, Friston K (2016) Computational phenotyping in psychiatry: a worked example. eNeuro 3(4)
57. Jobson DD, Hase Y, Clarkson AN, Kalaria RN (2021) The role of the medial prefrontal cortex in cognition, ageing and dementia. Brain Commun 3(3):fcab125
58. Buckner RL (2013) The cerebellum and cognitive function: 25 years of insight from anatomy and neuroimaging. Neuron 80(3):807–815
59. Caligiore D, Pezzulo G, Baldassarre G, Bostan AC, Strick PL, Doya K, Helmich RC, Dirkx M, Houk J, Jörntell H, Lago-Rodriguez A, Galea JM, Miall RC, Popa T, Kishore A, Verschure P, Zucca R, Herreros I (2017) Consensus paper: towards a systems-level view of cerebellar function: the interplay between cerebellum, basal ganglia, and cortex. Cerebellum 16:203–229
60. Jacobs HI, Hopkins DA, Mayrhofer HC, Bruner E, van Leeuwen FW, Raaijmakers W, Schmahmann JD (2018) The cerebellum in Alzheimer's disease: evaluating its role in cognitive decline. Brain 141(1):37–47
61. Bastos AM, Vezoli J, Fries P (2015) Communication through coherence with inter-areal delays. Curr Opin Neurobiol 31:173–180
62. Buzsáki G, Schomburg EW (2015) What does gamma coherence tell us about inter-regional neural communication? Nat Neurosci 18(4):484–489
63. Mauk M, Medina J, Nores W, Ohyama T (2000) Cerebellar function: coordination, learning or timing? Curr Biol 10(14):R522–R525
64. Adhikari A, Topiwala MA, Gordon JA (2010) Synchronized activity between the ventral hippocampus and the medial prefrontal cortex during anxiety. Neuron 65(2):257–269
65. Hein TP, Gong Z, Ivanova M, Fedele T, Nikulin V, Ruiz MH (2023) Anterior cingulate and medial prefrontal cortex oscillations underlie learning alterations in trait anxiety in humans. Commun Biol 6(1):271
66. Schultz W (2007) Multiple dopamine functions at different time courses. Annu Rev Neurosci 30:259–288

67. Schultz W (2016) Reward functions of the basal ganglia. J Neural Transm 123:679–693
68. Caligiore D, Mannella F, Baldassarre G (2019) Different dopaminergic dysfunctions underlying parkinsonian akinesia and tremor. Front Neurosci 13:550
69. Byrne GD, Hindmarsh AC (1975) A polyalgorithm for the numerical solution of ordinary differential equations. ACM Trans Math Softw (TOMS) 1(1):71–96

The manufacturer's authorised representative in the EU is Springer
Nature Customer Service Centre GmbH, Europaplatz 3, 69115 Heidelberg,
Germany. If you have any concerns regarding our products, please
contact ProductSafety@springernature.com

Printed and bound by CPI Group (UK) Ltd, Croydon, CR0 4YY
29/04/2026
02099543-0001